Welding Fabrication & Repair: Questions and Answers

Frank M. Marlow, P.E.

Illustrations by
Pamela Tallman

Industrial Press Inc.

Library of Congress Cataloging-in-Publication Data
Marlow, Frank M., PE
    Welding fabrication and repair: questions & answers / Frank M. Marlow: illustrations by Pamela Tallman.
    p. cm.
    ISBN 0-8311-3155-1
    1. Welding I. Title.
TS227 .M268 2002
671.5'2--dc21   2002025017

Industrial Press, Inc.
200 Madison Avenue
New York, NY 10016-4078

First Edition, May 2002

Copyright © 2002 by Industrial Press Inc., New York. Printed in the United States of America. All right reserved. This book, or any parts thereof, may not be reproduced, stored in a retrieval system, or transmitted in any form without the permission of the publisher.

STATEMENT OF NON-LIABILITY

The process of welding is inherently hazardous. Welding is often used to assemble buildings, structures, vehicles, and devices whose failure can lead to embarrassment, property damage, injury, or death. The author has carefully checked the information in this book, has had others with professional welding experience and credentials review it as well, and believes that it is correct and in agreement with welding industry practices and standards. However, the author cannot provide for every possibility, contingency, and the actions of others. If you decide to apply the information, procedures, and advice in this book, the author refuses to be held responsible for your actions. If you decide to perform welding, use common sense, have a competent person check your designs, the welding processes, and the completed welds themselves. Seek the help of people with professional welding inspection credentials to check critical welds until you gain experience. Do not attempt life-critical welds without competent on-the-spot assistance.

Printed in the United States of America

10   9   8   7   6   5   4

For Squeaker, Grey, and Freddie

Frank M. Marlow is a Registered Professional Engineer and holds a BA and BSEE from Lehigh, an MSEE from Northeastern, and an MBA from the University of Arizona. With a background in electronic circuit design, industrial power supplies and electrical safety, Marlow has worked for Avco, Boeing, Raytheon, Du Pont, and Emerson Electric. He has served as both Secretary and Treasurer of the Long Beach-Orange County AWS section. He is the coauthor of *Welding Essentials: Questions and Answers*, also published by Industrial Press, and is the author of *Machine Shop Essentials: Questions and Answers* (Metal Arts Press).

Contents

PREFACE		vii
ACKNOWLEDGEMENTS		viii
CHAPTER 1	FABRICATION BASICS	1
CHAPTER 2	BASIC BUILDING BLOCKS	27
CHAPTER 3	PIPE & TUBING	35
CHAPTER 4	BENDING & STRAIGHTENING	101
CHAPTER 5	VEHICLE WELDING	141
CHAPTER 6	WELDING PROBLEMS, SOLUTIONS & PRACTICES	149
CHAPTER 7	STRENGTH OF MATERIALS	251
CHAPTER 8	TOOLS, MATERIALS, SUPPLIES & INFORMATION	293
APPENDICES		299
GLOSSARY		301
INDEX		329
CREDITS		333

Preface

Beginning welding students learn how to make weld beads—and nothing else.

This book is for anyone who has basic welding knowledge and would now like to actually make something.

Welding Fabrication & Repair goes beyond the classroom, to real life practical applications such as vehicle frame repairs, building tables, rectangular and box frames, and brackets.

In *Welding Fabrication & Repair* you will learn:
- The ways weldors solve problems from building up a worn shaft with weld metal, to welding perforated steel screening onto frames, to extending the capacity of a welding machine.
- The checklist for designing a welded product.
- The most common pipe and vehicle welding methods.
- How weldments can replace castings.
- Leading suppliers of tools and equipment and their web sites.
- Flame bending and straightening tools and methods.
- Structural steel practices for joints, column splices, and bolting, and guidelines for sizing welds.
- The basics of the science of strength of materials.

Whether you're a serious welding student or a weekend weldor, I hope this book will help you develop your welding skills.

I would be happy to receive comments, corrections, and suggestions from readers via the Internet: weldbook@earthlink.net

HUNTINGTON BEACH, CALIFORNIA   FRANK MARLOW
MAY, 2002

Acknowledgements

The author wishes to express his special thanks and appreciation to The James F. Lincoln Arc Welding Foundation for granting permission to draw from two Foundation books: *Design of Weldments* and *Design of Welded Structures* both by Omer W. Blodgett. Without this material this book would not have been possible.

The author also acknowledges the assistance of the following individuals and organizations which made significant contributions to this book:

Andrew G. Kireta, Jr.
Applied Bolting Technology Products, Inc.
Assembly Technologies International, Inc.
Chicago Metal Rolled Products, Inc.
Copper Development Association Inc.
DE-STA-CO Industries
Di-Acro, Incorporated
Don Tallman
Haydon Bolts, Inc.
International Fastener Institute
Jeremy Long
Joseph Stafford
Mathey Dearman, Inc.
MK Products Inc.
Nucor Corporation, Fastener Division
Oxylance Corporation
Robert O'Con
Robert Weinstein
Van Sant Enterprises, Inc.
Victaulic Company of America
Vise-Grip® – American Tool Companies, Inc.
Walhonde Tools Inc.
Weldsale Company

Chapter 1

Fabrication Basics

*If you don't know where you are going,
you will probably end up somewhere else.*
—Laurence J. Peter

Introduction

Much more knowledge is required to make a welded item than just knowing how to produce a perfect weld bead. If you work in a company, you benefit from its established, step-by-step production procedures and the know-how of other employees. But, if you have just learned to weld and are welding on your own, you'll need some additional information. This is not surprising since basic welding courses focus on how to make welds, but do not go through the fabrication steps to make a complete welded item. Knowing how to perform each of these steps and how to use the tools that make these steps easier are critical to a successful welded fabrication.

This chapter provides this information along with a discussion of the proper welding environment, presents the typical steps in weld fabrication, and describes how each step is performed. It looks at tools to measure, mark, cut, and position materials for welding, including a variety of commercially popular brands of clamps, pliers, and welding fixtures. Also, several designs of welding and cutting tables and their accessories are detailed along with a look at the most common fabrication materials and how to cut them to size. Finally, the chapter details metal cleaning methods required before painting and the advantages of painting, powder coating, and anodizing.

The Appendix lists many leading companies that supply hand and power tools, and welding equipment. Also listed are companies that provide findings and fittings—hinges, wheels, axels, and bearings—to complete your welded project.

The Welding Work Environment

What are the principal factors in a safe and effective welding environment?

The welding area should:
- Be clean and comfortable to work in.
- Allow you to position the work to avoid welding on the floor unless absolutely necessary; you will not do you best work there.
- Be free of drafts on the work from fans, wind, windows, and doors, yet still have adequate change of air ventilation to reduce weld fume inhalation.
- Provide bright light; welding in sunlight is better than in dim light as the non-glowing parts of the weld show up better.
- Be between 70 to 80°F (21 to 27°C) because better welds will result than welds made in cold temperatures; however, acceptable welds can be made at ambient temperatures in the 40 to 50°F (4 to 10°C) range except where the weld specifications call for preheating.
- Have tools positioned within easy reach of the weldor.
- Be clear of combustibles, puddles, and tripping hazards.
- Provide all necessary personal safety equipment for the processes to be used.

Fabrication Steps

What steps are typical in weld fabrication?
- Getting or making a fabrication sketch or drawing.
- Developing a well thought out procedure.
- Gathering tools and materials.
- Making patterns, templates, and fixtures, if needed.
- Laying out and cutting material to be welded.
- Making edge preparations and cleaning metal areas to be welded.
- Making jigs or fixtures if needed.
- Positioning and clamping materials prior to welding.
- Tack welding assemblies, checking dimensions, setup, and squareness.
- Placing the final welds and assembling final fabrication.
- Painting the fabrication.

Tools

What are the most common fabrication hand tools and what are their applications?

- Builder's and torpedo levels—Use the larger builder's level whenever possible; it is more accurate and measures over a longer span; use the torpedo level wherever the builder's level won't fit.
- Framing, carpenter's, cabinet maker's, and combination squares—Use the largest square that fits the work. The combination square is convenient for layout of 45° corner cuts and parallel lines.

See Figure 1–1.

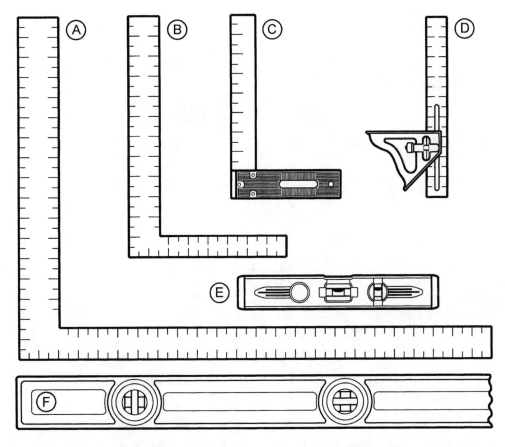

Figure 1–1. Common hand tools for weldors: (a) framing square, (b) carpenter's square, (c) cabinet maker's square, (d) combination square, (e) torpedo level, and (f) builder's level.

- Cold chisel and ball peen hammer—Handy for breaking tack welds when they must be repositioned; also useful for removing material between a series of drilled holes (chain drilling).
- Center Punch—Marks hole centers and cutting lines.
- Compass and dividers—For scribing circles or stepping off a series of equal intervals.
- Files—For bringing an oversized part down to exact dimension or removing a hazardous razor/burr edge.
- Hack saw—For slow, but accurate metal cutting.
- Tape measures—16- and 24-foot tapes are the most convenient sizes. Useful for measuring on curved surfaces too.
- Precision steel rules—Available in lengths from 6 to 72 inches (150 to 1000 mm).
- Protractor—For finding angles.
- Trammel points—These points fit on and adjust along a wood or metal beam and scribe circles or arcs with 20- to 40-foot (6 to 12 m) diameters. See Figure 1–2.

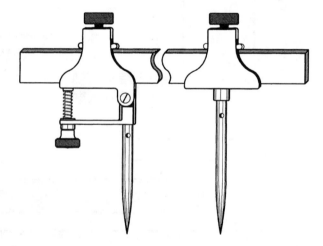

Figure 1–2. Trammel points for scribing large circles.

What are the most common methods for marking metal and their advantages?

- Chalk line snap
- Weldor's chalk, also called soapstone
- Ball point metal marker
- Single center punch mark or a line of punch marks
- White lead or silver lead pencil
- Felt-tip pen
- Aerosol spray paint
- Scriber on bare metal
- Scriber on layout fluid

Use weldor's chalk for marking rough dimensions or to indicate cutting lines that will hold up under cutting torch heat. A line of center punch marks can be more accurate and will also withstand torch heat. For very accurate layout lines, spray paint the metal in the area of the layout lines and use a scriber to scratch through the paint to make the layout lines. Alternatively, machinist's layout fluid (Dykem® is the major brand, and available in red or blue) can be used to make the scribed lines more visible. These lines will not hold up under torch heat, but can be essential to laying out non-torch cutting lines. A black felt-tip pen can also be used in place of spray paint or layout fluid to darken the metal and show up scriber lines. Do not use scribe marks to designate bend/fold lines since they will be stress raisers and the part will eventually fail along the scribed line. Metal markers are available that are a combination of a ball point pen tip mounted on the end of a squeeze tube, Figure 1–3. They put down a 1/16-inch width line, come in several colors, and are excellent for applying lettering to metals. The are rated at 700°F (370°C) so cannot be used for torch cutting lines. Note that some marking materials' residues may contaminate GTAW welds.

Figure 1–3. Ball point metal marker.

What are the most common power tools used in welding?

- Sawzall®-type reciprocating saw—Excellent for rough cuts of bar stock, shapes, pipe, and plate.
- Portaband®-type hand-held band saw—Capable of accurate cuts and following scribed lines; excellent for both pipe and tubing. Its throat is too small for most plate cutting.
- Electric drill, drill bits—For starter holes to begin oxyfuel cutting, holes to install hardware, and for chain drilling.
- Abrasive cutoff saw—Good for rod and pipe. Not good for shapes and tubing. Difficult to make accurate cuts.
- Portable grinder with abrasive and wire wheels and abrasive flap wheels—To remove mill scale, rust, and paint before welding. It is also good for smoothing rough edges and removing bad welds.
- Bench grinder/pedestal grinder with abrasive and wire wheels—Same as portable grinder, but here the operator holds the parts.

6   **CHAPTER 1**   **FABRICATION BASICS**

- Oxyfuel and plasma cutting torches—For rapid cutting of plate, shapes, and larger pipe.

See Figure 1–4.

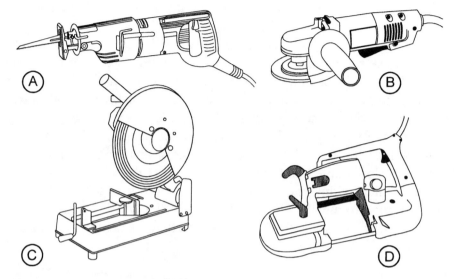

Figure 1–4. Common fabrication power tools: (a) reciprocating saw, (b) grinder, (c) abrasive cutoff saw, and (d) hand-held band saw.

What are the two most common problems a weldor faces in fabrication?
- First, holding the parts to be welded in the proper position to make the weld.
- Second, preventing weld-induced distortion.

Clamps and fixtures play a critical role in solving both these problems.

What kinds of clamps are available?
There are three general classes of clamps:
- General purpose clamps used in machine shops and carpentry.
- Clamps designed specifically for welding.
- Clamps designed to hold two (or more) specific items together for welding.

What type general purpose clamps are there?
- C-clamps.
- Bar clamps/pipe clamps.
- Kant Twist®-type heavy duty clamps.
- Come-alongs, also called cable pullers—Used to pull large/long members into position. These are particularly good on structural steel for pulling frames into square. See Figure 1–5.

Welding Fabrication & Repair

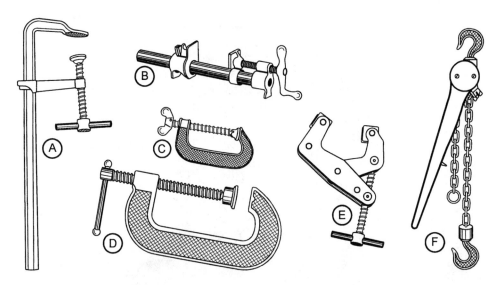

Figure 1–5. General purpose clamps used in welding: (a) bar clamp, (b) pipe clamp, (c & d) C-clamps, (e) Kant-Twist®-style clamp, and (f) come-along.

What clamps are designed specifically for welding operations?
- Bessey®-design welding clamps and pliers. See Figures 1–6 and 1–7.

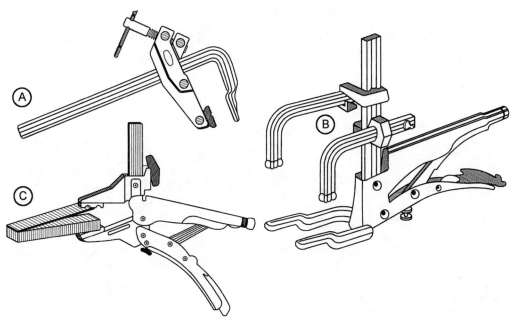

Figure 1–6. Bessey® heavy-duty welding clamps.

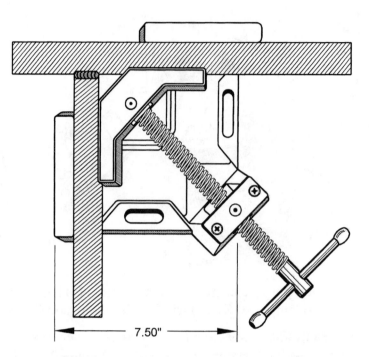

Figure 1–7. Bessey® welding corner clamps hold work firmly in position for welding materials at right angles.

- Angle clamps for tubing. These inexpensive, commercially available clamps can produce excellent results on light tubing, Figure 1–8.

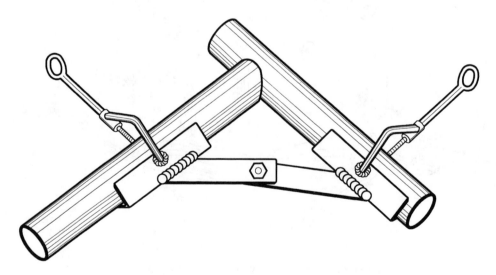

Figure 1–8. Angle clamps for welding tubing.

- Vise-grip® welding clamps and pliers—These tools are made for many applications: some pin two pieces of metal together, some hold a small part like a washer to a larger part, some hold two pipe pieces together, and others have a deep throat for an extra-long reach from the edges of the clamped material. See Figures 1–9 through 1–12.

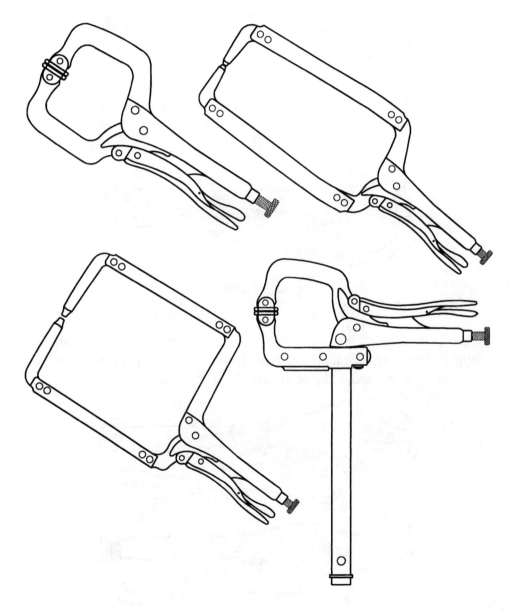

Figure 1–9. Vise Grip® welding clamps.

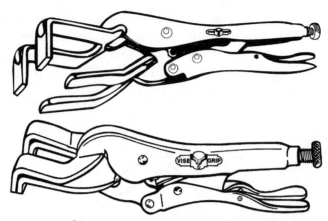

Figure 1–10. Vise-Grip® welding pliers, Model 9R are useful for making butt welds (upper) and Model 9AC holds thin strips of material together for edge welds (lower).

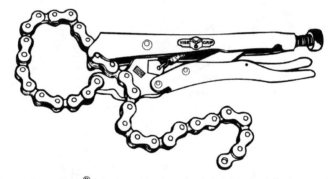

Figure 1–11. Vise-Grip® chain pliers, Model 20R holds irregularly shaped objects firmly against a pipe or tube. It can hold more than two objects together and add-ons to extend the chain are available.

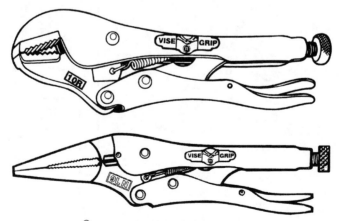

Figure 1–12. Vise-Grip® Model 10R and Model 9LN pliers are useful for a variety of welding tasks.

- DE-STA-CO Industries manual squeeze action clamps look more like pliers, but are really clamps. These clamps have either forged steel or aluminum frames. Smaller models provide 100 pounds (45 kg) of clamping force; the largest ones, 1200 pounds (540 kg) of clamping force. There are over thirty models providing a wide range of jaw openings and throat depth. They are popular because they can be set and removed quickly. On models with a quick release bar, just one hand is needed to open the clamp. A dozen or more clamps are often used to secure sheet metal panels or plates onto a frame for welding. Figure 1–13 shows the basic design; Figure 1–14 shows a clamp frame which provides a base for the customer's own clamping jaws.

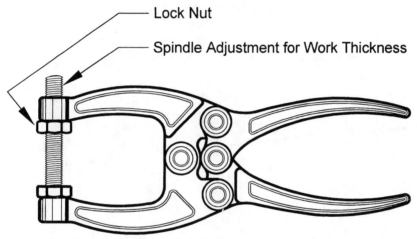

Figure 1–13. DE-STA-CO Industries squeeze action manual clamp useful for welding.

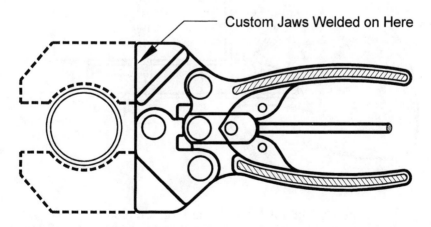

Figure 1–14. DE-STA-CO Industries squeeze action manual clamp for customer-supplied custom jaws.

What clamps are designed for welding specific items together such as large diameter pipe?
Figure 1–15 shows clamps designed to accurately position large diameter pipes for welding. Such clamps—those made for welding specific items—are called *fixtures*. A variety of commercial fixtures are available for pipe welding operations.

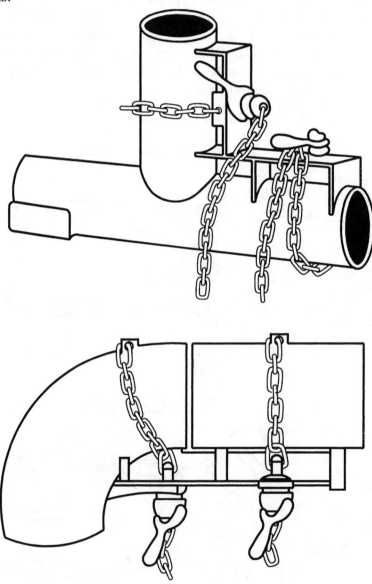

Figure 1–15. Clamps for welding large diameter pipe.

WELDING FABRICATION & REPAIR

When fabricating heavy plate, what other items are often useful in bringing adjacent plates into alignment prior to welding?
Puller clips and *bolts* can be helpful in moving plates into position. See Figure 1–16. These temporary bolts and brackets are removed usually by grinding after welding is completed.

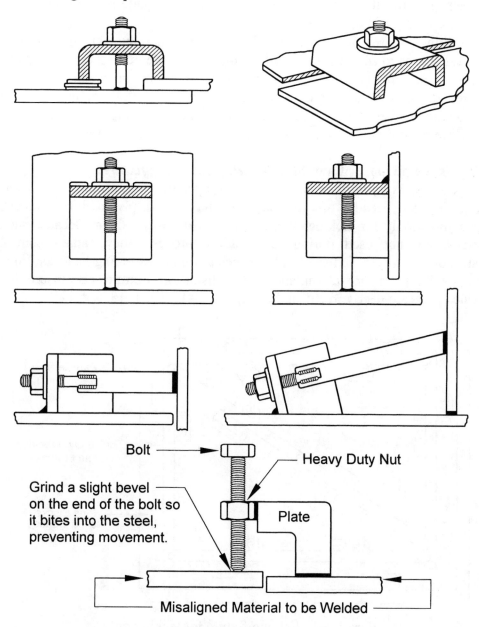

Figure 1–16. Puller clips and bolts provide the force needed to bring plates and panels into the proper position for welding.

Patterns and Fixtures

Why are patterns helpful?
There are several reasons:
- Many patterns for pipefitting, sheet metal, and other trades produce excellent results with little effort. Many of these patterns are available in handbooks or in commercial kits.
- Making patterns from paper, cardboard, or plywood lets you develop sizes and shapes to see how they will fit together before cutting metal. Patterns are especially helpful when copying a part from an existing item for repair. Besides showing the size and shape of a part, they can indicate hole positions, bend lines, reference points of meeting adjacent parts, placement of other parts on its surface, and weld bead locations.

What is the purpose of a welding/brazing/soldering *fixture*?
A fixture holds parts to be joined in the proper relationship to one another. Some fixtures, because of their materials (flakeboard, plywood, or lumber), are suitable only for tacking. Welding heat will destroy them. Because such fixtures are not used during the welding process, they cannot control distortion. However, fixtures made of metal *can* take welding heat; and they can position parts for welding and control distortion. Fixtures also insure that welded parts are nearly identical. See Figures 1–17 and 1–18.

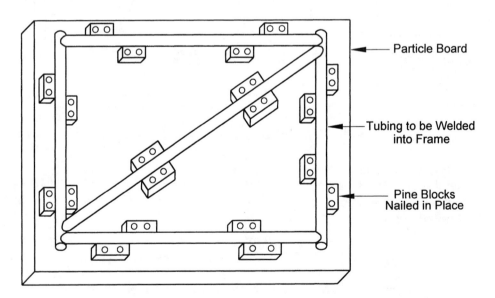

Figure 1–17. Welding fixture for tacking only.

WELDING FABRICATION & REPAIR   15

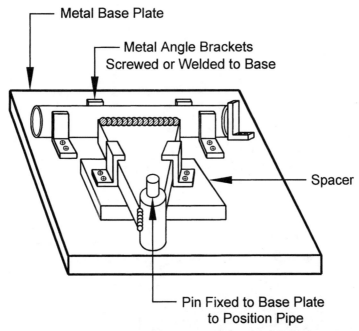

Figure 1–18. Metal welding fixture to withstand welding heat and control distortion.

- Adding quick-release clamps like those in Figure 1–19 makes a metal welding fixture like the one in Figure 1–18 suitable for production. These clamps can exert hundreds of pounds of pressure to hold components in place for welding, yet let finished parts come out of the fixture easily. Some quick-release clamps are designed to be screwed in place, Figure 1–19 (a), (b), & (e), while others for heavy-duty service are welded to the fixture base (c). For large fixtures, where one person cannot easily and quickly actuate manual clamps, or where there are many clamps spaced around the edges of a large piece of work, compressed-air actuated clamps are available (d).

What are three ways to approach building a welding fixture?
- Assemble one complete part to the dimensions of its plans, then build the fixture to fit the part.
- Cut all the parts for the welded item. Fit them together, then build a fixture that fits and secures the parts.
- Build a fixture from the plans for the item, then build the first part.

Depending on the complexity of the item, the thoroughness of its drawings and the tolerances to be held, one approach will likely be better than another. There is no hard and fast rule; each case must be considered separately.

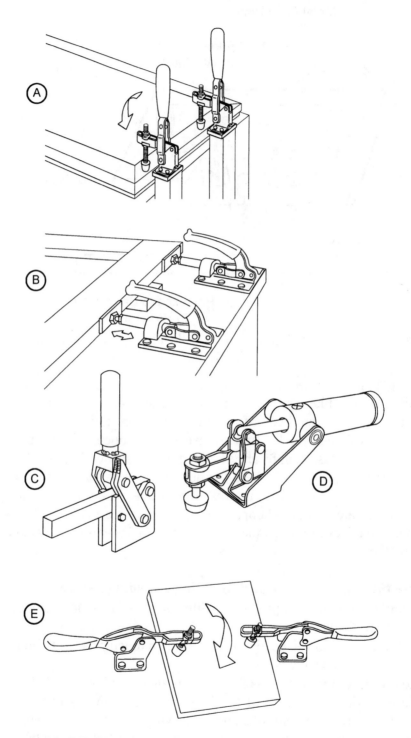

Figure 1–19. DE-STA-CO Industries quick-release clamps for fixtures.

WELDING FABRICATION & REPAIR 17

Why should you have a welding table with a steel top similar to the one in Figure 1–20?

A welding table places the work at a comfortable height and allows the weldors to concentrate on their work, rather than their discomfort. A welding table allows some welds to be made sitting down. Also, it provides a stable, flat surface to position and clamp work prior to welding. In some applications the work itself may be tack-welded to the table. Later these weld-tack beads can be ground off. Finally, the lower shelf provides a convenient place to lay tools.

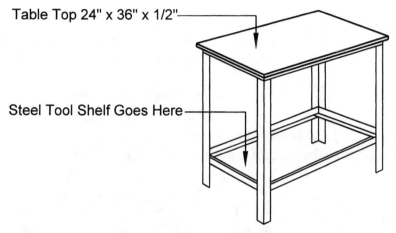

Figure 1–20. Welding table.

What table design should be used when flame or plasma cutting?

It should be similar to the table in Figure 1–20, but have a grilled top as in Figure 1–21. The grill will suffer less damage from the cutting flame than a solid steel top.

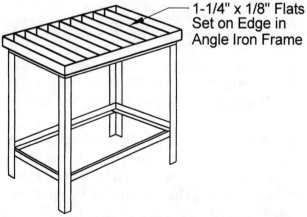

Figure 1–21. Table for flame or plasma cutting.

What other items are helpful to have with a welding or cutting table?
Several fire bricks to get the work off the table when oxyfuel welding, brazing, soldering, or welding. Also helpful are a metal bucket or tub of water for cooling work and a fire extinguisher.

Is there a way to reduce flame cutting damage to the cutting table?
Yes, use cast iron cutting *pyramids*. Pyramids are made in various sizes and their bottom slots match the width of the cutting table cross bar supports, so they fit onto the bars and support the work. See Figure 1–22. Pyramids are used to:
- Prevent cutting flame damage to the cutting table support bars.
- Eliminate blowback that results in cutting tip contamination and loss of cutting action when the cutting flame crosses over cutting table support bars.
- Stop the work from welding to cutting table support bars.
- Reduce slag cleanup as it blows slag clear of the workpiece.
- Provide solid, level support for the work.
- Support parts so they do not sag while they are being cut.

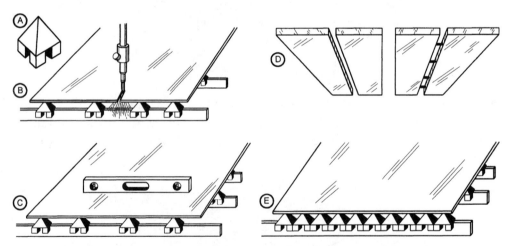

Figure 1–22. (a) Cast iron cutting pyramid, (b) protecting cross supports, (c) leveling and supporting the work, (d) reducing slag where flame crosses support bars, (e) full support for parts by moving pyramids closer together.

You have a project too large to fit on the welding table shown in Figure 1–20 and will require extensive clamping to position the parts for welding. What kind of welding table would be appropriate?
A welding *platen* and its related clamps would be helpful and often essential for building complex weldments, Figures 1–23 and 1–24. Small platens are

2 × 3 ft (0.6 × 1.0 m) across and weigh about 475 pounds (215 kg); the large platens are 5 × 10 ft (1.5 × 3.0 m) across and weigh 4080 pounds (1850 kg). The top surface of the platens are Blanchard ground to assure flatness. The corner edges provide bolting faces so the platens can be ganged to provide a larger work area. For small and medium size weldments, platens are mounted on legs bringing the platen work surface 32 inches (81 cm) from the floor. For very heavy jobs, the platens sit directly on the floor and can have forklifts or cranes drive onto them to handle the workpieces.

Platens have a rectangular array of 1¾ inch (3.7 cm) square holes cast into their surface which provide convenient places to mount a variety of clamps and fixtures. Clamps install from the top of the platen; therefore no access to the platen underside is needed. Clamps drop into the square holes and rotate 1/8 turn so the four small pockets or depressions on the underside edges of each square hole capture the edges of the clamp so it can be tightened. Many different style clamps are available and any clamp will fit into any hole. See Figure 1–24.

For some tasks one or more platens are mounted vertically on top of horizontal platens. On other jobs, each end of the workpiece may sit on a different platen.

Platens provide these advantages:
- A flat, level, and dimensionally stable surface for layout, bending, straightening or fabrication.
- With cutting pyramids to prevent damage, platens are useful for oxyfuel or plasma cutting.
- Tooling holes provide an ideal path for downdraft fume extraction systems.
- Complex weldments can quickly be clamped up, eliminating the need for specialized tooling.
- The many different types of clamps provide excellent restraint to minimize distortion when welding.

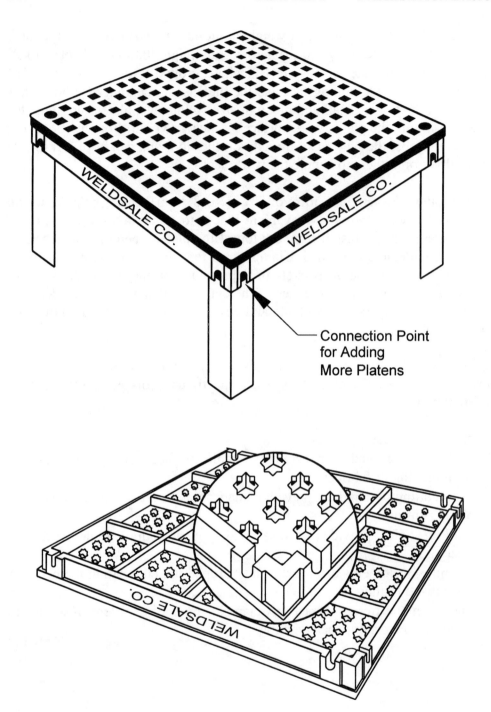

Figure 1–23. Cast iron welding platen mounted on legs (top), underside view of same platen showing detail of bolt pockets in square holes (bottom).

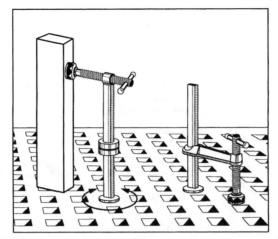

Figure 1–24. Welding platen clamps: horizontal clamp (left) and vertical clamp (right).

Steel Products

What are the most common carbon steel products used in welding fabrication?

Low carbon, hot rolled solid shapes, sheet goods, plate, pipe, and tubing are the most often used fabrication materials. Large steel distributors stock a wide variety of shapes and sizes. See Figure 1–25. Additional sizes and shapes of steel as well as other materials like alloy steels, stainless steel, brass, copper, and bronze are also available on order. These materials are substantially more expensive than carbon steel.

When planning projects, remember that one size of tubing (either round, square, or rectangular) is dimensioned to *telescope* or slide smoothly into the next larger size. This can be very helpful and a design shortcut.

Large steel distributors usually have a variety of remnant material that is sold by the pound. A tape measure and calipers can also help you determine if a particular remnant will be useful. This can be both economical and convenient for many projects. Bring work gloves to handle the greasy and sharp remnants.

Usually for a small charge, or often at no charge, the distributor will cut the material to make it easier to transport since many products come in 20 foot lengths. With good planning, cuts made on the distributor's huge shears or band saws can save you a lot of cutting time, particularly on heavy plate goods.

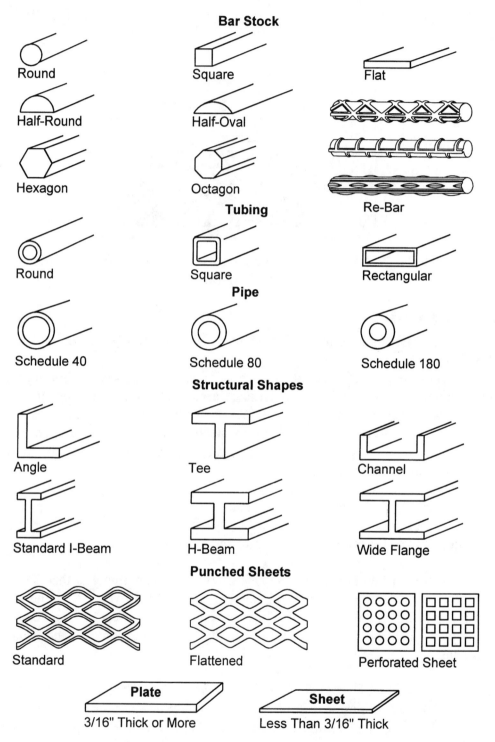

Figure 1–25. Carbon steel fabricating materials.

WELDING FABRICATION & REPAIR

In what sizes and size increments are plates, sheets, bars and rounds available?
See Table 1–1.

Product	Maximum Thickness	Width or Diameter	Incremental Dimensions
Sheet Steel	>3/16" thickness	36-84"	Even numbered gauges are most common; other gauges on special order.
Plate Steel	≤3/16" thickness	> 8" to 60"	1/32" up to 1/2" thickness, then 1/16" up to 1" thickness, then 1/8" up to 3" thickness, and 1/4" above 3" thickness
Rectangular Bars		≥ 8"	Thickness: 1/16" increments, but usually 1/8" increments. Width: 1/4" increments
Round Bars		≥ 8"	Diameter: 1/8" increments

Table 1–1. Product size and increment availability.

What other hardware items beside steel goods do weldors commonly use in fabrications?
- Nuts—By welding a nut over an existing hole, we can add threads without tapping them; this facilitates adding leveling jacks, adjustment members, and clamps.
- Bolts—These can be used for leveling jacks, axels, swivel points, and locating pins. Bolts can also be cut to provide just the threaded portion for threaded studs, or just the rod section.
- Allthread—This is rod stock threaded end-to-end and useful where clamping or positioning action is needed.
- Hinges—There are three main hinge designs:
  - Leaf hinges for welding—These hinges are not plated and have no screw holes.
  - Cylindrical weld hinges—These are made in a wide variety of sizes and can support heavy loads.
  - Piano hinges—Provide continuous support along a door or cover.
  See Figure 1–26.
- Casters and Wheels—These are better purchased than shop-made.

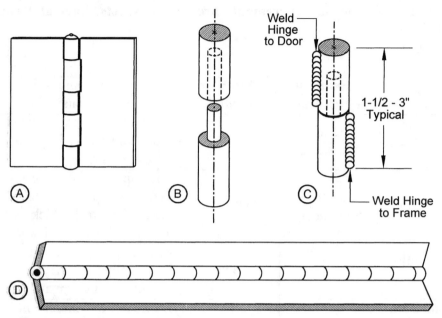

Figure 1–26. Weld-on hinges: (a) Leaf hinge has no plating or screw holes, (b) heavy-duty cylindrical hinges as supplied, (c) cylindrical hinges when installed, and (d) piano hinge.

Cleaning

What metal cleaning preparation should be done for GMAW and FCAW?

These processes are fairly tolerant of mill scale and small amounts of rust and paint so it is possible to make good welds on most steel rolled goods—flats and shapes—as they come from the mill. However, the metal must not be greasy and for this reason most hollow products like pipe, tubing, and hollow rectangular shapes that are shipped from the factory well oiled must be degreased before welding.

What are typical degreasing agents?

Household cleaners like Simple Green® or Formula 409® All Purpose Cleaner will work; industrial degreasers like denatured alcohol and acetone can also be used. Paint stores, metal supply houses, hardware stores, and pool supply stores often carry phosphoric acid (dilute in 4 to 10 parts water). These stores may carry tri-sodium phosphate (TSP), also a good cleaner. Do not clean hollow steel shapes too far in advance of welding or they will rust. Do not use compressed air to dry them off as this will re-introduce oil contamination from the compressor. Use a plastic bristle brush or a stainless steel brush since copper, brass, or aluminum brushes will contaminate the weld.

Welding Fabrication & Repair

In addition to using one of the above degreasing agents, what other standard cleaning steps are usually used?
- Grind, wire brush with a grinder, use flap wheels, or emery cloth to remove all mill scale, rust, paint, and dirt and get down to fresh metal, Figure 1–27.

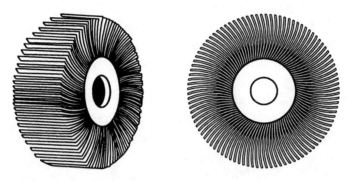

Figure 1–27. Flap wheels for removing rust and scale.

- Wipe cleaned weld area with alcohol or acetone to remove residual grease.
- Avoid getting your fingerprints on the area just cleaned.
- Remember not to cross-contaminate your wire brush, emery cloth, and flap wheels by using the same ones on both steel and stainless steel. Hint: If you will be working with both steel and stainless steel, paint the handles of stainless steel brushes red for use on steel, and green for use on stainless. This prevents cross-contamination.

Protective Metal Finishes

Why are protective finishes particularly important for welded items?
Preparation for welding removes mill scale, grease, and paint, thus exposing fresh, bare metal to the atmosphere, an ideal condition for rapid corrosion. This is particularly true for most steels and for aluminum in a salt atmosphere. A protective finish will prevent corrosion and enhance the part's appearance.

What are the most common protective finishes for welded products and their advantages/disadvantages?
- Painting for all metals—No specialized equipment is needed, but spray painting may be best for complex shapes to reduce labor expense. Several coats may be needed to make the item weatherproof.
- A red Rust-oleum® brand primer and two more finish coats will provide at least five years of rust-free service outdoors. In general, products which are supplied in aerosol cans are less durable than those supplied in conventional cans.

- Powder coating for steel and aluminum—Provides a durable, professional-looking surface with many colors and surface textures available. May be nearly as inexpensive as painting for complex shapes as it is sprayed on. Holds up well outdoors.
- Anodizing for aluminum only—This coating is durable and very thin, typically from 0.5 to 6 thousandths of an inch (0.013 to 0.154 mm). Colors are available but tend to fade in the sun. Red is the least stable color, black is much more stable. A non-colored, clear anodizing is the most stable. This is not a do-it-yourself process; leave it to specialists.

When should protective coatings be applied?
Protective coatings should go on promptly after welding (or final surface prep) so the metal does not have a chance to react with the atmosphere. Ideally only a few hours should elapse.

Chapter 2

Basic Building Blocks

*Millions say the apple fell,
but Newton was the one to ask why.*
—Bernard Baruch

Introduction
This chapter contains step-by-step instructions for making rectangular frames, putting legs on a table, and building a three-dimensional solid frame.

Rectangular Frames
What are two ways to make angle iron corners for a rectangular frame?
Mitering and notching. See Figure 2–1. A beginner might find notching easier, since there is no need make 45° angle cuts. After welding, grinding, and painting, frames made by either method will look equally good.

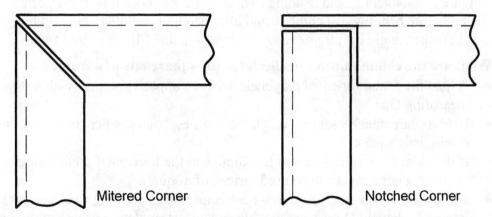

Figure 2–1. Mitered (left) and notched (right) frame corners.

What other method can be used to make rectangular frames with angle iron?
See Figure 2–2. This approach lends itself to production work and the best results are achieved when a notching machine is available. The extra material between the 45° cuts provides the material to go *around* the outside of the corner of the frame when the bend is made, Figure 2–2 (a). Begin by setting the bend allowance gap to the half the thickness of the angle iron and go from there. Additional length must be added to the frame for the bend allowance gap metal.

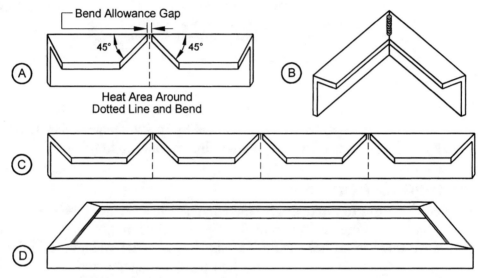

Figure 2–2. Notching and bending to make a single-piece frame: (a) Corner detail before bending, (b) corner detail after bending, welding, and grinding, (c) notched angle iron frame ready for bending, and (d) completed frame.

What are the common ways to check for the squareness of a frame?
- Adjust the frame for equal diagonals between opposite corners with a steel measuring tape.
- Use a carpenter's square on large frames; on smaller ones use a machinist's square.
- If the sides of the frame are to be plumb and the horizontal sections are to be level, a large level can be used instead of a square.
- If a square is too small, use a 3-4-5 triangle when welding two members into an L-shape: (1) measure off four units (feet, meters, or some multiple of the same) on one leg, and (2) measure off three units on the other leg, (3) then adjust the hypotenuse, the longest side of the triangle, to measure 5 units. This procedure makes a perfect right triangle. See Figure 2–3.

WELDING FABRICATION & REPAIR

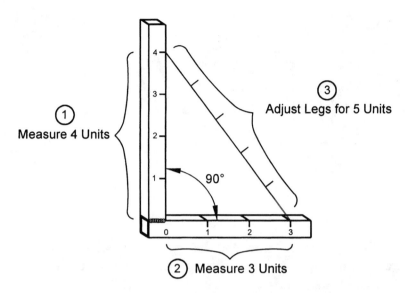

Figure 2–3. Using a 3-4-5 triangle to set members at a right angle.

When working with larger members, how should the length of the members be adjusted?
By reducing the overall length of the frame sides to allow for additional joint root spacing. If you don't, the finished frame will be oversized.

What steps can improve the chances of a frame being welded square?
In *decreasing* order of effectiveness:
- Use a rigid fixture to secure the members in place while welding them.
- Clamp members to a steel table, then weld them.
- Use a fixture to hold parts for tacking, then weld the tacked parts *outside* the fixture. This fixture can be as simple as a sheet of plywood with wood blocks fixed to it to hold the work in place while the tack welds are made.
- Use Bessy®-type corner clamps. Hint: Begin by tack welding each of the corners together using the clamp and checking for squareness after each tack. Bend the frame back into squareness if needed. If the tack welds are not too large, you will be able to straighten the frame by hand. Begin final welds at *opposite* corners. Warning: If you make a single weld one corner at a time using a corner clamp, you will not like the results. The final two corner pieces will not meet.
- Use magnetic corner tools—These are effective only for light sheet metal as they lack the strength to resist weld-induced distortion forces even with light angle iron, Figure 2–4.

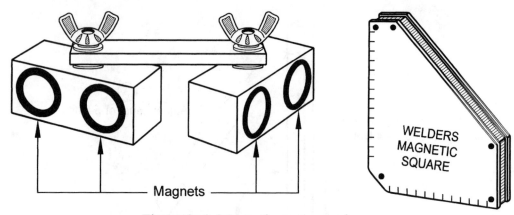

Figure 2–4. Magnetic corner tools.

- In the field, or with large and heavy members, lay the members up squarely on the floor or ground using shims to get them level, and check the diagonal measurements for squareness. Tack the corners together, and then weld them. Check to see that the frame is both square and level after making each tack, and bend members back to squareness and flatness *before* making the next tack or weld. If your tacks are too big, large (and unobtainable) forces may be needed to bring the frame back into square. *Warning: Welding directly on concrete can cause it to explode violently. Avoid this hazard, by keeping the work off the floor using shims or spacers.*
- Weld one corner, and then weld the opposite corner before welding the remaining two corners. Weld the *same* relative corner or side position in the exact *same* sequence on all four corner joints: weld all outside faces, then all top corners, finally all bottom faces. Make each weld in the same relative direction: from the outside of the frame to the inside or vice versa. Also, give each weld a moment to cool before making the next one.

You have welded a rectangular frame of angle iron (not rectangular tubing) and it does *not* lie flat. Now what?

Follow the steps in Figure 2–5 showing how to bend the horizontal face of the frame to flatten it. Use an open-end wrench, or fabricate a tool of your own.

WELDING FABRICATION & REPAIR

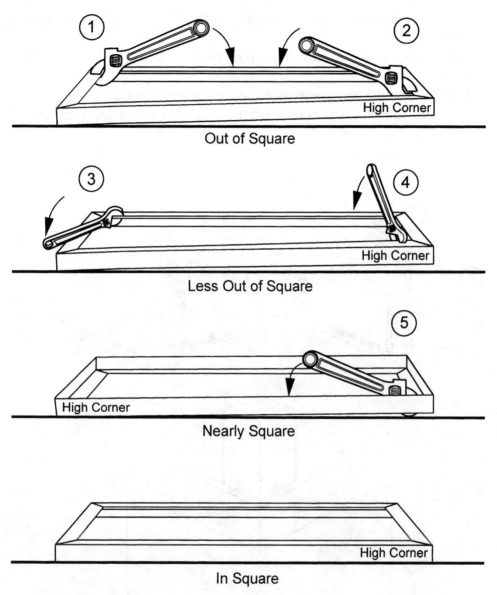

Figure 2–5. Method of adjusting an angle iron frame to lie flat.

Mounting Table Legs

You are making a rectangular table frame and have completed the top. How are the legs attached so they are welded on square?

- Put the table frame upside down on a flat surface.
- Use two clamps to lightly secure a leg to both sides of the frame corner, Figure 2–6 (a).
- With a carpenter's square, adjust this leg so it is perpendicular to the

frame and using a length of steel or wood use two clamps to brace the leg to bring it into square.
- Repeat this squaring/bracing/clamping for the other right angle, Figure 2–6 (b).
- Fully tighten the two clamps holding the leg to the frame.
- Re-check squareness in both directions, adjusting as needed, then weld the leg to the frame, Figure 2–6 (c).
- Repeat for each leg.

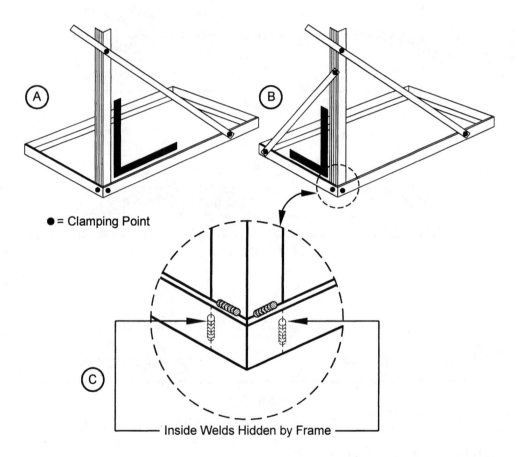

Figure 2–6. Welding table legs on square.

Box Frames
You are making a rectangular box frame. How should you do this?
- Make the upper and lower frames as described previously.
- Secure the four verticals to the lower frame as if they were table legs, Figure 2–6.
- Place the completed upper frame over the legs.
- Make whatever compromises and adjustments are needed to the verticals to make them meet the upper frame. Some tweaking may be necessary. Note that all opposite diagonals, such as dotted line X-Y, will be the same length in a *rectangular* box.
- Clamp the legs to the upper frame.
- Tack all joints, check for squareness and fit, then weld all joints.

This method will work equally well with angle iron or rectangular tubing. See Figure 2–7.

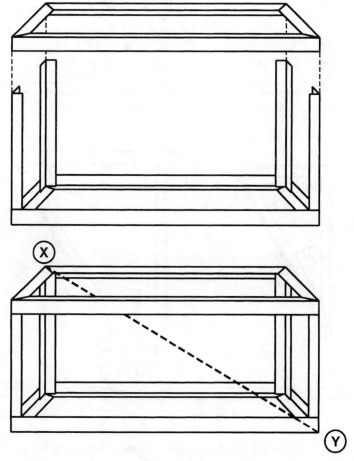

Figure 2–7. Making a box frame.

Angle Brackets

What are the two most common designs for steel angle support brackets?
They are 45-45-90-degree triangles and 30-60-90-degree triangles, Figure 2–8. These brackets are strong, easy to make, and can be made from three lengths of angle, welded at each corner, or if a notching machine is available they can be made from a single length of angle bent twice and welded at one corner. Steel angle brackets are used to support pipes, cables, electrical equipment and machinery.

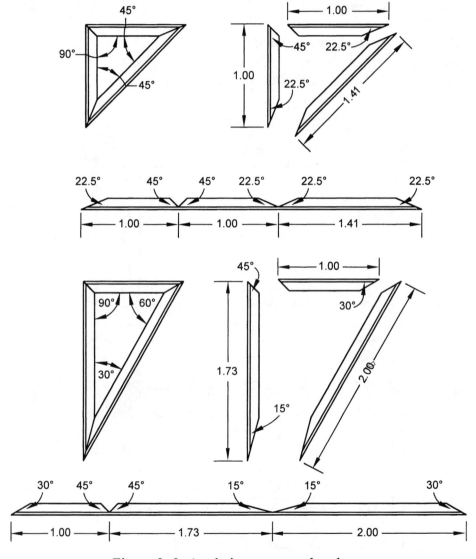

Figure 2–8. Angle iron support brackets.

Chapter 3

Pipe & Tubing

Crank—a man with a new idea until it succeeds.
—Mark Twain

Introduction
Even in the age of microprocessors, pipe and tubing play an indispensable role in our everyday lives. Steel pipe brings oil to the surface, moves it to refineries via pipelines, and works in heat exchangers, catalytic cracking units and other refinery equipment to make finished petroleum products. Then more pipelines move refined products like gasoline and diesel fuel to end-users. Boiler tubes, steam lines, cooling lines, instrumentation, and control lines in power stations bring us comfort and convenience. Water and natural gas distribution systems bring these essentials to our homes, plants, and offices. Semiconductor manufacturing requires welded stainless steel tubing in its chip lines. Food processing plants, dairies, and breweries require miles of welded stainless steel pipe and tubing. Literally hundreds of combinations of size, construction, and alloys are needed to make modern life possible. Pipe and tubing are also used in building furniture, aircraft, racing car frames, and structural elements of building and machinery. Welding and brazing are the most common methods to join structural pipe and tube.

This chapter begins by looking at the different types of pipe and tube, then presents the two approaches to making tubular products, and the advantages of welded pipe and tubing joints. It also details every step of the pipe welding process and examines the types of joints and fittings.

Pipe vs. Tubing
What is the difference between *pipe* and *tubing*?
Pipe usually has a much thicker wall than tubing. These thicker walls permit pipe to accept threads and still have enough steel remaining to provide adequate wall strength to handle fluid pressure. Because tubing has thinner walls, it cannot be threaded.

Another difference is that pipe from 1/8- to 12-inches (6 to 300 mm) diameter is specified by its *inside* diameter; pipe 14-inches (355 mm) diameter and larger is specified by its *outside* diameter. See Table 3–1. Tubing is always specified by its *outside* diameter and wall thickness. Since not all tubing is round, its shape (round, square, or rectangular) must also be specified.

Pipe & Tubing Sizing

Why does the *pipe size* diameter differ from the pipe's actual diameter? For example, why does a 2-inch pipe have an actual diameter of 2.375 inches?

The size of pipe is identified by its *nominal pipe size (NPS)*. For pipes between 1/8- and 12-inches nominal size, the outside diameter (OD) was originally selected so that the inside diameter was equal to the nominal size for pipes of standard wall thickness of the times. This is no longer true with the changes in metals and manufacturing processes, but the nominal size and standard OD continue in use as a trade standard.

How is pipe size specified in the metric system?

Each nominal pipe size in inch-pound units has its metric equivalent called the *diameter nominal (DN)*. Both "25 mm" and "2 inch" are names the trade gives to pipes of 2-3/8 inch OD. Fittings, flanges, couplings, valves, as well as other piping components, are interchangeable in the two measurement systems thanks to the work of the International Standards Organization. The first (NPS) and second (DN) columns in Table 3–1 show these equivalences.

How is the wall thickness of pipe and of tubing specified?

Pipe wall thickness is specified by its *schedule*. There are schedules from schedule 10 through schedule 180. Schedule 40 pipe is considered the *standard* pipe or wall thickness. In general, if you ask for a steel pipe of a specific nominal diameter and do not call for a specific schedule, you will get schedule 40 pipe. Those pipes in schedules lower than schedule 40, have thinner walls; those pipes in schedules above schedule 40 have thicker walls and can sustain greater pressures. Not all wall thicknesses (or schedules) are manufactured in all pipe diameters. Table 3–1 shows the available sizes and schedules.

For a given pipe material, each schedule number represents a safe working pressure range. Pipe wall thickness must increase as the pipe's diameter increases to sustain the increased forces on its walls. All pipes with a given schedule number are rated for the same pressure range. By calling for a

Welding Fabrication & Repair

certain schedule pipe, the users are automatically assured the pipe will operate properly within a given pressure range and they do not have to be concerned with what the pipe's wall thickness must be for *each* diameter of pipe used. See Table 3–1 and Figure 3–1 for comparisons of pipe schedules.

Nominal		O.D.	PIPE SCHEDULES WALL THICKNESS										
NPS Inches	DN mm	Inches	5	10	20	30	40 & Std	60	80 & E.S.	100	120	140	160	180 & D.E.S.
1/8	6	.405	.035	.049			.068		.095					
1/4	8	.540	.049	.065			.088		.119					
3/8	10	.675	.049	.065			.091		.126					
1/2	15	.840	.065	.083			.109		.147				.187	.294
3/4	20	1.050	.065	.083			.113		.154				.218	.308
1	25	1.315	.065	.109			.133		.179				.250	.358
1 1/4	32	1.660	.065	.109			.140		.191				.250	.382
1 1/2	40	1.900	.065	.109			.145		.200				.281	.400
2	50	2.375	.065	.109			.154		.218				.343	.436
2 1/2	65	2.875	.083	.120			.203		.276				.375	.552
3	80	3.500	.083	.120			.216		.300				.437	.600
3 1/2	90	4.000	.083	.120			.226		.318					.636
4	100	4.500	.083	.120			.237	.281	.337		.437		.531	.674
4 1/2	115	5.000					.247		.355					.710
5	125	5.563	.109	.134			.258		.375		.500		.625	.750
6	150	6.625	.109	.134			.280		.432		.562		.718	.864
7		7.625					.301		.500					.875
8	200	8.625	.109	.148	.250	.277	.322	.406	.500	.593	.718	.812	.906	.875
9		9.625					.342		.500					
10	250	10.750	.134	.165	.250	.307	.365	.500	.500	.718	.843	1.000	1.125	
11		11.750					.375		.500					
12	300	12.750	.165	.180	.250	.330	.375	.562	.500	.843	1.000	1.125	1.312	
14	350	14.000		.250	.312	.375	.375	.593	.500	.937	1.093	1.250	1.406	
16	400	16.000		.250	.312	.375	.375	.656	.500	1.031	1.218	1.437	1.593	
18	450	18.000		.250	.312	.437	.375	.750	.500	1.156	1.375	1.562	1.781	
20	500	20.000		.250	.375	.500	.375	.812	.500	1.280	1.500	1.750	1.968	
24	600	24.000		.250	.375	.562	.375	.968	.500	1.531	1.812	2.062	2.343	
26		26.000		.312	.500		.375		.500					
28	700	28.000		312	.500	.625	.375							
30	750	30.000		.312	.500	.625	.375		.500					
32	800	32.000		.312	.500	.625	.375		.500					
34		34.00		.312	.500	.625	.375							
36	900	36.000		.312		.625	.375		.500					

Table 3–1. Standard pipe size and wall thickness.

To add to the confusion, the terms Standard (Std.), Extra-Strong (E.S.) and Double Extra-Strong (D.E.S.) are also used to describe pipe wall thickness. Table 3–1 and Figure 3–1 show how these terms fit in with the pipe schedules.

Tubing wall thickness is specified in inches (mm) or *Manufacturers' Sheet Metal Gauge*, Table 3–2. In general, decimal inches or Sheet Metal Gauge number are used for square or rectangular steel tubing for structural or mechanical applications. Decimal inches are used for round steel and stainless steel tubing designed for fluid transport. Round tubing is usually stocked in outside diameter increments of 1/16 inch.

There are more than a dozen different thickness gauge schemes (or tables) listed in engineering handbooks. Some are obsolete, some are not used in the U.S., some apply only to ferrous or only to non-ferrous metals, and others only to wire diameter. There are many opportunities for confusion. If in doubt which table to use, specify pipe wall thickness in decimal inches or mm.

What are the main types of tubular goods?
See Figure 3–2. Combinations of size, wall thickness, and alloys run into the hundreds. This table does not include many other common pipe and tubing products not of particular interest to weldors—for example, electrical conduit and thin wall electrical tubing, cast iron soil pipe, cement-based domestic water supply pipe, and the variety of plastic pipes used in home and industry.

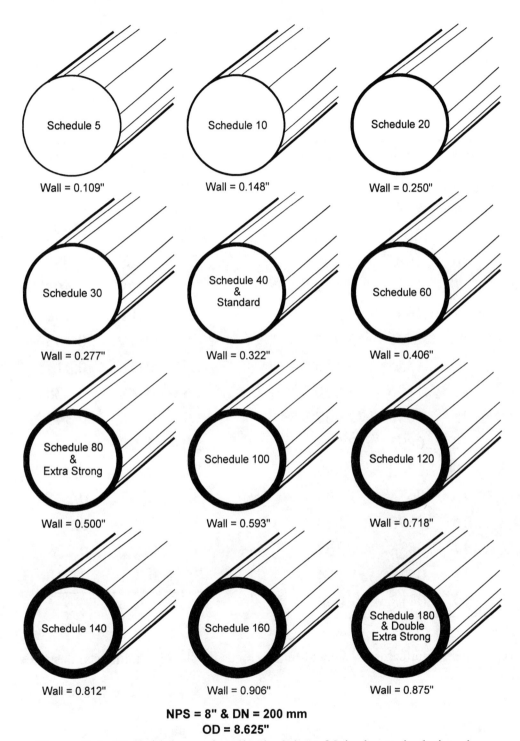

Figure 3–1. Wall thickness for all schedules of 8-inch nominal pipe size.

Gauge	(inches)	(mm)	Gauge	(inches)	(mm)
3	.2391	6.073	20	.0359	0.912
4	.2242	5.695	21	.0329	0.836
5	.2092	5.314	22	.0299	0.759
6	.1943	4.935	23	.0269	0.683
7	.1793	4.554	24	.0239	0.607
8	.1644	4.176	25	.0209	0.531
9	.1495	3.797	26	.0179	0.455
10	.1345	3.416	27	.0164	0.417
11	.1196	3.030	28	.0149	0.378
12	.1046	2.657	29	.0135	0.343
13	.0897	2.278	30	.0120	0.305
14	.0747	1.897	31	.0105	0.267
15	.0673	1.709	32	.0097	0.246
16	.0598	1.519	33	.0090	0.229
17	.0538	1.367	34	.0082	0.208
18	.0478	1.214	35	.0075	0.191
19	.0418	1.062			

Table 3–2. Manufacturers' Sheet Metal Gauges for steel including stainless steel.

WELDING FABRICATION & REPAIR

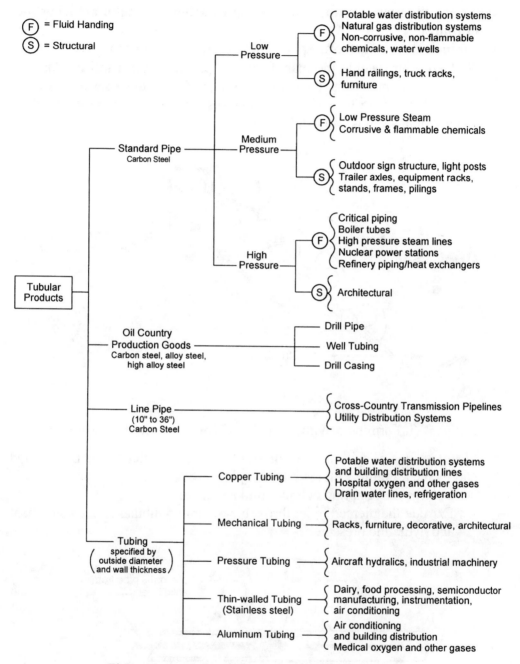

Figure 3–2. Common types of pipe and tubing.

What are the most common process approaches for manufacturing pipe and tubing?

Two principal manufacturing approaches are *welded seam* and *seamless*.

- Welded seam begins the manufacturing process with a roll of flat steel which is cut into strips, then, with the help of heat and a series of rollers, the steel strips are formed into a tubular shape. To complete the pipe the longitudinal seam is welded, Figure 3–3.

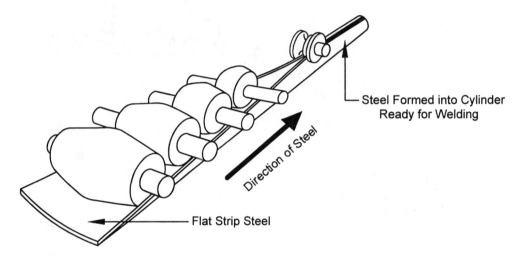

Figure 3–3. Forming a cylinder from strip metal for welded-seam pipe.

- Seamless begins with a billet or solid cylinder of metal which is heated and pierced, Figure 3–4, then, to complete the process:
  - Enlarge the resulting cylinder into a pipe, or
  - Extrude the pierced billet through a die into a tubular shape using heat and hydraulic pressure.

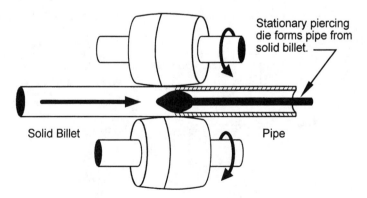

Figure 3–4. Forming a pipe from a billet by piercing.

WELDING FABRICATION & REPAIR

In the welded approach, what processes are used to weld up the longitudinal seam?
There are several:
- The oldest method for closing the pipe seam is a high-speed forge welding process called continuous welding (CW). After the pipe is heated and formed into a cylinder, it is drawn through a set of dies at high temperature and pressure. The dies squeeze the longitudinal seam together, making a forge weld. Pipe from 1/8- to 4-inch (3 to 100 mm) diameter is made this way.
- There are three similar processes using electrical energy to make a resistance weld along the longitudinal seam. The choice of process depends on the diameter of the pipe, the wall thickness, and the production rate. These methods use radio-frequency induction, sliding contacts, or rollers to transfer electrical energy to heat the weld.
- DSAW (double submerged arc welding) is used in heavy wall applications.
- GTAW (gas tungsten arc welding) is often used on seamed stainless tubing.
- Electron beam and laser processes are also used, but are less common.
- The quality of the product, tightness of inside and outside diameter tolerances, and the kind of metal determine the process.

Section I - Pipe

Pipe Welding Applications and Advantages

When is pipe welding used?
Welding is used for the majority of large diameter pipe joints since it is more cost effective than using threaded joints. All sizes of pipe and structural pipe joints are usually welded. Although the majority of pipe is used for fluid transport, pipe is widely used for structural and load-carrying applications.

What are the advantages of pipe welding?
- Cost savings from eliminating joining fittings
- Cost savings from using thinner walled pipe when threads are not used
- Labor savings over threaded pipe
- Reduction of fluid flow resistance from smoother interior joint surfaces versus those of screwed fittings from turbulence reduction, Figure 3-5
- Ease of repair
- Higher strength, vibration resistance, more leak-resistant than threaded joints

- Easier to apply insulation with bumps of joint fittings eliminated
- More pipes can be installed in less space with welding than with screwed fittings as "swing" room must be made with threaded piping

Studies have shown that the labor time to thread or weld 1¼-inch (32 mm) size pipe are about the same, but as pipe diameter increases, welding gains a strong advantage, so that on 6-inch (150 mm) diameter pipe, welding takes only about half the time as threading. Also, the equipment and personnel needed to move around a pipe length to thread it, position it, and make it up, become more and more of a problem as pipe diameter increases. With welding, the pipe only need be brought into alignment once.

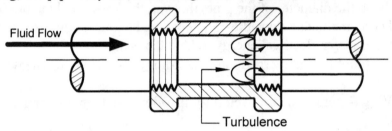

Figure 3–5. Turbulence generated by step where pipe meets fitting.

Piping Codes

What codes govern pressure piping in the US?

Most pressure piping is governed by the ASME piping codes of B31 with the welding procedures generally in accordance with *Section IX of the ASME Pressure Vessel Code*. While the transmission and distribution systems for gas and liquid petroleum products are governed by *API Standard 1104*, this document is in general agreement with B31.8 of the ASME Codes. See Table 3–3.

ASME Code Section	Application
B31.1	Power piping
B31.2	Fuel gas piping
B31.3	Chemical/petroleum plant piping
B31.4	Liquid transportation systems for hydrocarbons, liquid petroleum gas, anhydrous ammonia and alcohols
B31.5	Refrigeration piping
B31.8	Gas transmission and distribution systems
B31.9	Building services piping
B31.11	Slurry transport systems

Table 3–3. ASME codes for pressure piping.

Pipe Welding Procedures

What is a welding procedure specification (WPS)?
A WPS sets down the required welding variables (or conditions) for a specific application to assure repeatability by properly trained weldors. The ultimate objective is to design a WPS and through testing, prove that the weld produced by the WPS has the characteristics to meet the requirements of its specific application and relevant governing code. Some codes include prequalified WPSs, so if the user follows the WPS recipe, welds made with it meet the code requirements. When a code provides a WPS, the user does not have to qualify the weld design. Some codes do not provide a WPS, and state only the testing parameters the weld must pass to be acceptable. This leaves the user to develop his own WPS for the application.

What welding variables are described in the WPS?
The welding variables in the WPS are:
- Base material
- Base material thickness/pipe diameter
- Electrical characteristics (polarity, current, voltage, travel speed, wire feed speed, mode of metal transfer, electrode size)
- Filler metal
- Preheat conditions
- Postweld heat treatment
- Shielding gas
- Weld type (groove, fillet)
- Welding position(s)
- Welding process

Welding Qualification

Who can make critical welds?
Critical welds, where life and property will be threatened should a weld fail, must be made by a weldor who has proven he can make a satisfactory weld following the relevant WPS. Such a weldor is *qualified* to perform a weld to that specific WPS.

Parts of the Weld

What are the terms used to describe the parts of a groove weld?
- Leg or size of weld
- Face
- Toe
- Reinforcement

- Root penetration

See Figure 3–6.

Most pipe welds are groove welds, but some are fillet welds. Fillet welds are used on socketed joints, on pipe saddles (when one pipe joins another without fittings between them), some flanges, and where a pipe joins a flat surface.

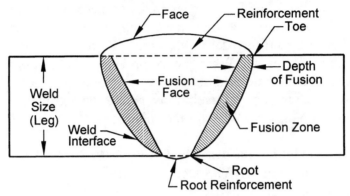

Figure 3–6. Parts of a groove weld.

What are the terms used to describe the parts of a fillet weld?

- Depth of fusion
- Face
- Weld size or leg
- Reinforcement
- Root
- Toe
- Weld fusion face and zone
- Weld interface
- Actual throat and theoretical throat

See Figure 3–7.

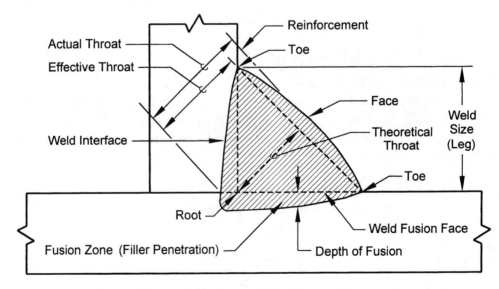

Figure 3–7. Parts of the fillet weld.

Pipe Welding Positions
What are the welding positions for pipe?
See Figure 3–8. Note the difference between welding positions A and B: In position A, the 1G horizontal rolled position, the pipe may be *rotated* about its longitudinal axis to provide access to any part of the weld joint. In practice, an assistant (or motor-driven work positioner) slowly rotates the pipe to keep the welding at the top of the pipe and in the flat position. Some weldors prefer to weld just slightly down from the top center of the pipe. This is much easier for the weldor than having to weld up or down hill. In position B, the pipe is fixed and *cannot* rotate forcing the weldor to weld up and down hill as well as overhead.

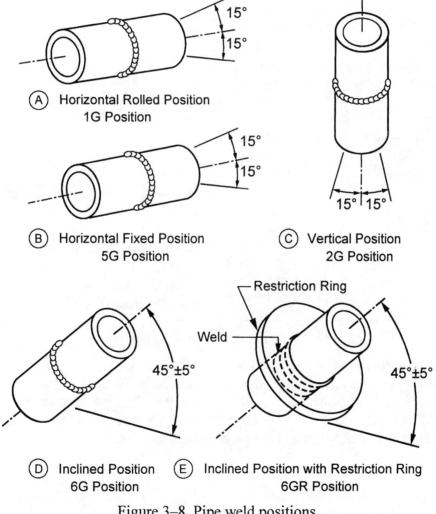

Figure 3–8. Pipe weld positions.

Pipe Welding Overview

What are the typical steps in installing a section of welded pipe?
The typical steps are:
- Cut pipe to length.
- Prepare pipe ends by cleaning, grinding, brushing, and beveling.
- Preheat the pipe to bring it up to proper temperature.
- Insert consumable backing rings and spacers, if any are needed.
- Position the pipe against the next length of pipe or fitting to which it will be welded so there is proper and equal root space between them.
- When using GTAW and *not* using an insert, protect the back side of the weld from atmospheric contamination with inert gas, usually argon.
- Apply tack welds to secure pipes and grind the tack welds.
- Perform the weld.
- Perform postweld heat treatment, if required.
- Perform visual and NDT examination.

Not all steps must be performed for every joint. We will examine each of these steps in more detail below.

Pipe Cutting

How is carbon steel pipe cut?
See Table 3–4.

Type Pipe	Method of Cutting
Steel pipe 3" (80 mm) diameter or smaller	Manual wheeled pipe cutter, or Wheeled pipe cutter operating in powered threading machine, or Portable bandsaw, or Oxyfuel cutting
Steel pipe over 3" (80 mm) diameter	Oxyfuel cutting, or Abrasive cut-off saw

Table 3–4. Cutting methods for carbon steel pipe.

How should steel pipe be cut with an oxyfuel torch?
In cutting small diameter pipe, the torch remains in the vertical position at all times. First, a cut from the 12 o'clock to the 9 o'clock position is made, and then another cut from the 12 o'clock to the 3 o'clock position is made. The pipe is rotated a half-turn and the process repeated. See Figure 3–9.

Welding Fabrication & Repair

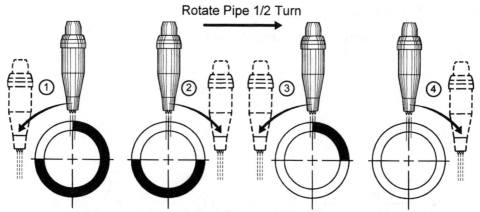

Figure 3–9. Cutting small diameter pipe with an oxyfuel torch.

Cutting large diameter pipe requires that the torch axis remain pointed at the center of the pipe and angled to make a bevel. Because larger diameter pipes are heavy, the pipe is not rotated, instead the pipe remains stationary and the torch rotates around the pipe's circumference, Figure 3–10.

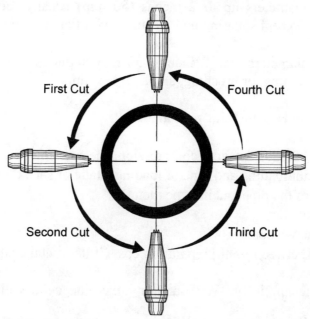

Figure 3–10. Cutting large diameter pipe with an oxyfuel torch.

Edge Preparation

What are the parts of an edge preparation?
See Figure 3–11.

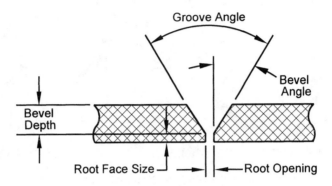

Figure 3–11. Parts of V-groove joint edge preparation.

What types of edge preparations are used for pipe welding?
- Pipe with diameters up to 2 inches (50 mm) usually requires no edge preparation except squaring and cleaning with flap abrasive wheel or wire brush.
- Pipe with diameters over 2 inches (50 mm) requires end preparation by cleaning, grinding, or brushing, and then beveling.

What factors determine joint preparation?
The factors are:
- Kind of joint loads: tension, compression, shear, or torsion.
- Level of joint loading and type of loading: static or shock loads.
- Thickness of the pipe and its type metal.
- Welding position.
- Skills of the weldors.
- Trade-offs between joint preparation costs, filler metal costs, and welding labor costs.

In practice, the applicable WPS and/or governing code will specify joint preparation.

What are the most common types of joint preparation?
See Figure 3–12.

WELDING FABRICATION & REPAIR

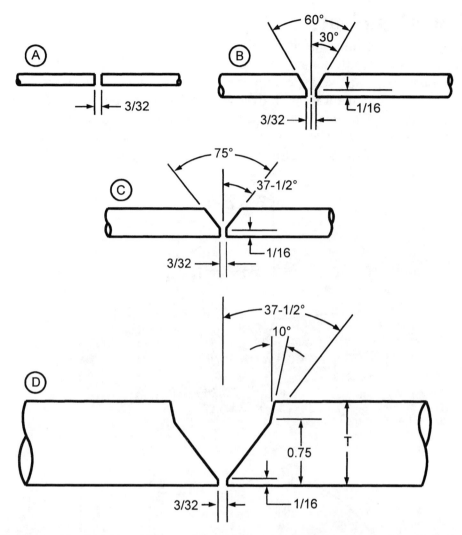

Figure 3–12. Common pipe welding edge preparations. Typical applications are (a) small-diameter pipe, (b) pipelines, (c) pressure pipes, (d) heavy-wall pipe.

How are weld joint preparations applied?

Simple bevels are usually applied with an oxyacetylene torch, air- or motor-driven cutting tool. Many weldors prefer to grind the oxyacetylene bevel smooth *before* starting to weld. Applying the entire bevel with a grinder is too slow for production work and is usually limited to repairs. Although oxyacetylene cutting torches may be hand-held, there are fixtures that clamp onto the pipe end and provide a track to keep the torch in alignment as it rotates around the pipe. These may be hand- or motor-driven. Most large diameter pipe comes with factory-applied edge preparation.

Pipe Alignment

How is pipe alignment accomplished?

There are many pipe alignment tools designed for this purpose. Some are lever operated, some screw clamp operated, and others air- or hydraulically-driven. Most alignment tools fit over the outside of the pipes, but, in pipeline work, they may go inside the pipes too. Many alignment devices not only align the pipe for correct root spacing, but also apply force to the pipes to correct for out-of-roundness. See Figures 3–13 through 3–17.

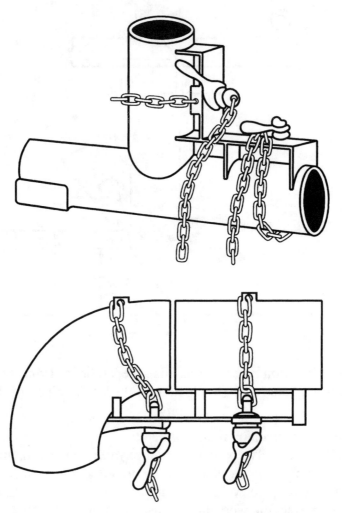

Figure 3–13. Chain clamps for large pipe alignment.

How is the Mathey Dearman rim clamp in Figure 3–14 used to align two pipe lengths for welding?

After edge bevels have been applied, the pipes are laid end-to-end with the appropriate root spacing. The rim clamp is applied over the joint so the stabilizer bars and reforming screws go onto one length of pipe (the left one in Figure 3–15) and the jack bar goes onto the other pipe section (the right one in Figure 3–15). The stabilizer bars and jack screws are adjusted to clamp the pipes into coaxial alignment. Then, reforming screws above high spots on the left hand pipe are adjusted so the pipe surfaces on both pipes around the weld bevels are even. The root pass, hot pass, and filler pass can be made from the right side of the rim clamp in Figure 3–15. The jackbars can be backed off and pinned back for grinding access and for the cover pass. All welding can be performed without removing the rim clamp from the right hand side of the clamp.

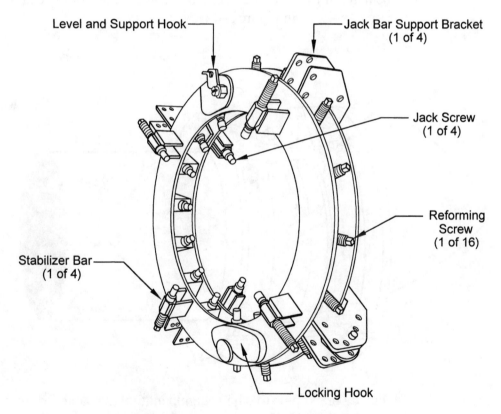

Figure 3–14. Mathey Dearman brand rim-type reforming clamps are available for 4- to 72-inch diameter pipe.

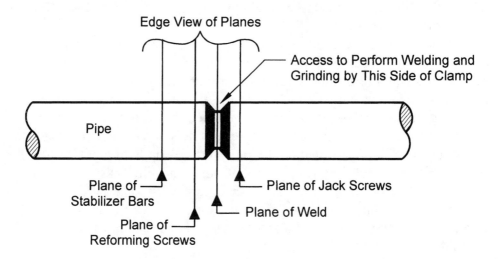

Figure 3–15. Schematic of Mathey Dearman brand rim-type reforming clamp in Figure 3–14.

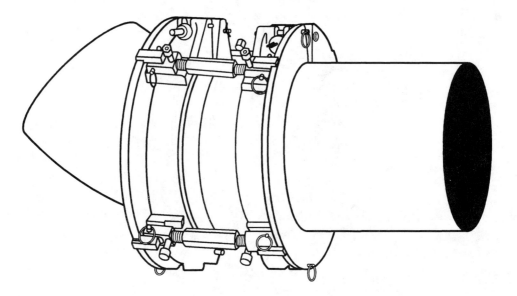

Figure 3–16. Walhonde brand pipe to elbow alignment tool available for from 4- to 20-inch diameter pipe.

WELDING FABRICATION & REPAIR 55

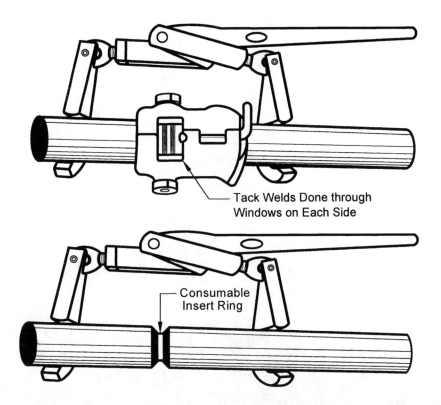

Figure 3–17. Walhonde brand "Boilermaker Kit" (above) and "Boomer" for use with consumable insert and chill ring (below). Both tools are used to align superheaters, reheats, economizers, crossovers, and penthouse header stubs.

Consumable Insert Rings

When are consumable insert rings used?
They are used when making GTAW welds with both manual and automatic equipment. They are used for both piping and tubing. Consumable inserts are useful because they protect the inside (or backside) of the weld from the atmosphere and take the place of filling the pipe or tube with inert gas to do the same thing. Because of the cost of the inert gas, or the difficulty of confining the gas within the section of the pipe under weld (particularly in large diameter pipes), weld inserts are a useful alternative.

For what other reasons are consumable insert rings used?
Consumable insert rings offer other benefits:
- Helping align the pipe ends and automatically set the root gap.
- Reducing root bead cracking due to notch effect.

What consumable insert ring designs are available?

See Figure 3–18 for typical insert ring cross-sections. There are also combination insert ring and backing ring designs, Figure 3–19.

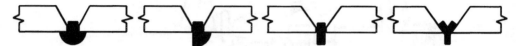

Figure 3–18. Consumable insert rings in cross-section.

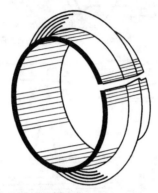

Figure 3–19. Combination insert and backing ring.

Preheating Pipe for Welding

Why and how is preheating of pipe performed prior to welding?

Preheating pipe is done with rose-bud oxyfuel torches, but resistance and induction heating methods are used too. They offer more precise temperature control than a torch.

Preheating is done to assure that excessive heat energy put into the weld is not drawn away to the surrounding cold metal, preventing proper fusion and creating a defective weld. Having the weld metal and the surrounding metal at 100 to 300°F (40 to 150°C) *before* welding begins ensures welding heat is not drained away to the surrounding cold metal.

Additionally, because preheating reduces temperature differentials between the weld zone and the surrounding metal, stresses from expansion are reduced, and the tendency of cracking and distortion are reduced. Preheating also helps entrapped gases, especially hydrogen, escape from the weld metal. The welding process, electrode, and pipe metal will indicate through the WPS what prewelding temperature is needed. Temperature-indicating crayons, portable pyrometers, or thermocouples determine when the metal is at temperature.

Tacking

What is *tacking* and why is it performed?
Tacking is the application of temporary welds in the joint root area. These welds hold the pipes to be joined in alignment for welding. Usually four ¾- to 2-inch (20 to 50 mm) welds at 90° intervals are used. When the actual joint welding is performed these tacks are fused into the weld joint.

Where should the tacks be placed?
Tack welds should be placed away from weld bead starts and stops. We want to place the tacks so the torch will encounter them only after the arc has fully stabilized and is generating maximum heat. This insures the tack weld metal is fully remelted into the weld bead. Beginning a weld on a tack may lead to discontinuities. See Figure 3–20.

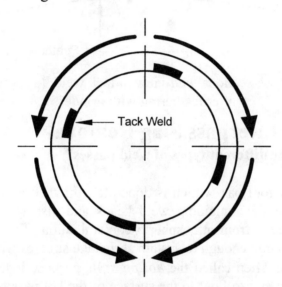

Figure 3–20. Place tack welds out of starts and stops of subsequent weld passes.

In critical welds, what additional steps must be taken after tack welding and before beginning the root pass?
Each end of each tack weld must be ground down or "feathered" with a narrow grinding wheel to assure complete fusion with the root pass weld metal. See Figure 3–21.

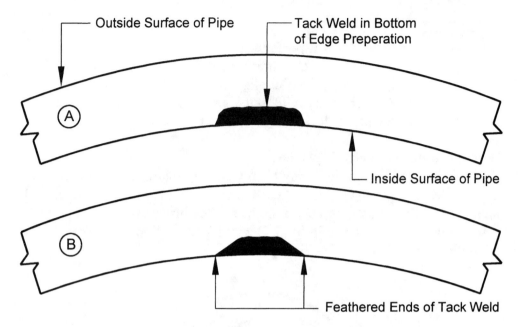

Figure 3–21. Cross-sectional end-view of tack weld: (a) as welded and (b) after feathering with grinder.

Welding & Interpass Slag Removal

What are the four different types of weld passes?
They are the:
- *Initial pass* or root pass which is done after the tack welds. It provides a solid basis for the remaining weld beads and also shields the back of subsequent beads from atmospheric contamination. The size of the root bead must be large enough to sustain shrinkage stresses without cracking.
- *Second pass* is often called the *hot pass*. Its purpose is to get slag inside the root pass metal to float to the surface of the hot pass bead for removal by grinding or chipping. Also, the hot pass insures complete edge fusion between the root pass metal and the base metal. Less metal is actually deposited during the hot pass than on subsequent filler passes. The heat energy (welding current) going into the hot pass is usually set slightly higher than the remaining passes to accomplish this. Note that many weldors speak of "burning out the slag" on the hot pass. This is not true. The steel of the pipe will burn before the slag will. The hot pass, when properly done, merely makes the slag accessible for removal.
- *Filler pass(es)* or intermediate pass(es) are those weld beads applied after the tacks and before the cover pass.
- *Cover pass* should have a neat appearance and a slight crown (1/16 inch or 2.5 mm) and not more than 1/8 inch (5 mm) to reinforce the weld.

WELDING FABRICATION & REPAIR

Each bead should start and stop at a different location from the bead over which it is deposited to minimize the accumulation of weak spots. This sequence of root pass, hot pass, filler passes and cover pass, when properly done, will produce a *full penetration weld* that is just as strong as the pipe itself, possibly stronger. Not all pipe joints need such a weld either because they operate at a relatively low pressure, or because their failure is not critical. See Figure 3–22.

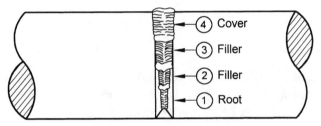

Figure 3–22. Sequential passes in a weld.

When should grinding and chipping be done?

Grinding should be done on the root pass. This will remove slag, reduce the high crown of the root pass, and leave a groove. This assures the hot pass will be on sound metal and leave no voids. This step is essential to achieve a full penetration weld. See Figure 3–23. Chipping or wire brushing should be done on all remaining passes to remove flux.

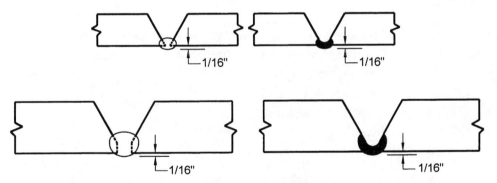

Figure 3–23. Root pass weld for light- (upper row) and heavy-walled pipe (lower row): Before (left column) and after grinding (right column).

Why is excessive reinforcement of the cover pass undesirable?

Too much of a crown on the cover pass makes the entire weld zone a constrictor ring and creates excessive stress between the weld and pipe metal as each responds to temperature changes. See Figure 3–24.

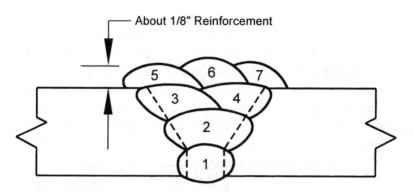

Figure 3–24. Proper cover pass reinforcement.

What temperature should be maintained on the weld and surrounding metal after preheat and welding begins on multiple-pass welds?

The temperature between weld passes, the *interpass temperature*, should not be allowed to fall below the preheat temperature for the same reasons preheating is valuable. There may also be a maximum temperature requirement in the WPS. If the heat buildup is excessive, wait until the temperature falls and then proceed.

Typical Pipe Welding Processes

What determines the welding procedure used?
The Welding Procedure Specification (WPS) indicates what weld procedure is required.

What are typical welding processes for carbon steel pipe?
See Table 3–5. The process abbreviations are ASW nomenclature:
- FCAW–flux core arc welding
- GMAW–gas metal arc welding
- GTAW–gas tungsten arc welding
- OFW–oxyfuel welding
- SMAW–shielded metal arc welding

WELDING FABRICATION & REPAIR

Pipe Diameter English/ Metric	Wall Thickness	Welding Position(s)	Welding Process	Row Ref. No.
4 in or less (100 mm or less)	Standard	All	OFW	❶
Up to 12 in (Up to 300 mm)	Standard	All All	SMAW GMAW	❷
4 to 12 inches (100 to 300 mm)	Standard & Heavy	All	GTAW(root)/SMAW	❸
4 in and larger (100 mm and larger)	Thin & Standard	All downhill All up or downhill All up or downhill	SMAW FCAW GMAW	❹
4 in and larger (100 mm and larger)	All	All uphill Rolled (1G) Rolled (1G) Rolled (1G)	SMAW SAW GMAW(root)/SAW GMAW(root)/FCAW	❺
All sizes	All	All	FCAW GTAW(root)/FCAW GTAW(root)/GMAW	❻

Table 3–5. Typical carbon steel pipe welding processes.

When is OFW (oxyfuel welding) used for pipe welding?
While OFW (row ❶) can be used to join steel pipe of any size and wall thickness given enough welding time, it is relatively slow compared with modern electrode-based processes which are simply faster and more cost-effective for most applications. However, OFW is still widely used in natural gas distribution systems, some water distribution systems, and in the repair of smaller process piping. See Figure 3–12 for typical joints. In these applications on smaller diameter pipe, no edge beveling is needed. On pipe 2 inches (50 mm) and smaller, no additional edge preparation is needed except squaring the pipe ends.

What are the relative advantages of using SMAW (shielded metal arc welding) versus GMAW (gas metal arc welding) for welding pipe up to 12 inches (300 mm) diameter?
See row ❷ of Table 3–5. SMAW is less sensitive to wind than GMAW and GMAW carries the additional cost of shielding gas. GMAW has the advantage of speed over SMAW; it can deposit a lot of metal rapidly without stopping to change electrodes. SMAW may also require more weldor skill.

Why is uphill welding usually used with SMAW on larger diameter pipes?
Although downhill welding is a faster way to deposit metal, uphill welding provides easier control for a large weld puddle as in row ❺.

Why is the rolled (1G) or flat position used for SAW (submerged arc welding) and FCAW (flux core arc welding) for larger pipes?
The resulting weld pool is more easily controlled in the flat position. See row ❺.

Are any pipe welding processes fully automatic (*AU* in AWS nomenclature)?
Yes, there are SAW (submerged arc welding), GMAW, and FCAW welding systems which are similar in design to the fully automatic GTAW orbital welder described in the welding of tubing in Section II. They can perform the processes in rows ❺ and ❻.

Postweld Heat Treatment
What is the purpose of postweld heat treatment?
Postweld heat treatment is performed to relieve stress. It is similar to the processes of annealing and normalizing, but done at lower temperature. The metal is never raised above its critical temperature. The weld metal is gradually raised to temperatures in the range of 1050 to 1200°F (566 to 649°C), held at temperature for about one hour per inch of wall thickness, and then cooled at the rate of 300 to 350°F (260 to 316°C) per hour.

Resistance or induction heaters with automatic temperature control and recording systems provide the heat. Times, temperatures, and heating/cooling rates must be followed precisely and are detailed in the treatment specifications. Postweld heat treatment *cannot* be performed with an oxyfuel torch.

In addition to relieving stresses, corrosion, caustic embrittlement, and shock-load resistance may be improved.

Weld Visual and Non-Destructive Testing (NDT)

What types of weld testing are performed on critical pipe applications?
Most welds are given a visual inspection, and those in critical applications get x-ray inspection also.

Pipe Joint Fittings

On many pipe welds 2 inches (100 mm) diameter and under, socketed joints are used. Why?
Figure 3–25 shows a socketed weld joint. When used with smaller pipes, socketed joints offer the advantage that the pipes only need be cut off squarely on the ends, no beveling is required, and the fillet weld is relatively easy to apply. These fittings are usually forgings.

As pipe goes from a small-diameter to a large-diameter, what change must be made in fitting design?
Smaller pipes usually use socketed or sleeved fittings, Figure 3–25 and 3–26. Both socketed and sleeved joints use fillet welds. Larger pipes use forged steel fittings, Figure 3–27. Pipes too large for factory-forged fittings usually have joints fabricated from the pipe material itself.

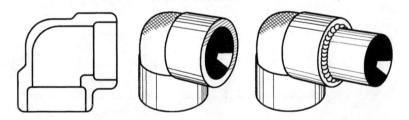

Figure 3–25. Socketed and forged steel 90° elbow in cross-section (left), before welding (center), and with fillet weld on pipe (right).

Is there another way to join smaller diameter pipes without socketed fittings or beveled end preparation?
Yes, use a lap collar and a fillet weld. On many low-pressure applications a single pass does the job. This kind of joint is used in many municipal water distribution systems, Figure 3–26.

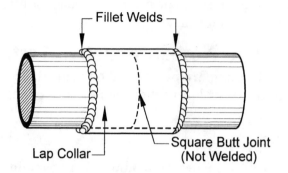

Figure 3–26. Lap-collar with fillet weld joint.

When the application has higher pressures than a lap collar or socket fitting will handle, what fitting is used?
Forged steel with factory-beveled ends, see Figure 3–27.

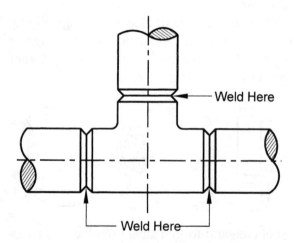

Figure 3–27. Forged steel fitting on beveled pipe ends for high-pressure applications.

What other styles of forged steel fittings are available?
Fittings for almost every imaginable application are available, Figure 3–28.

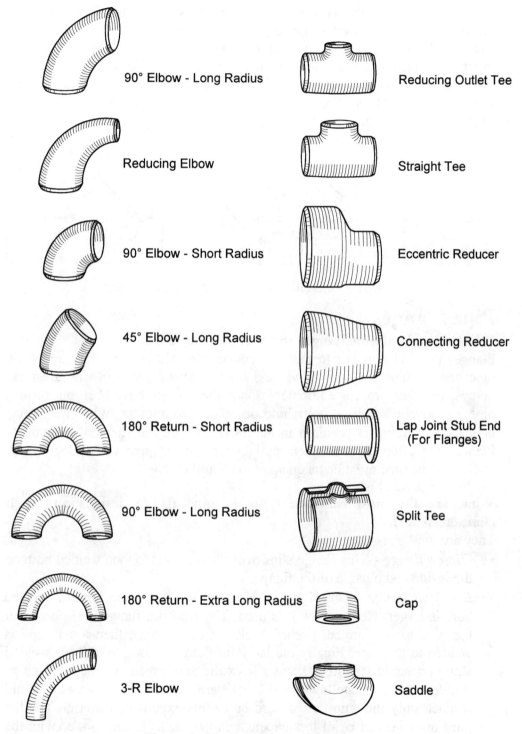

Figure 3–28. Forged steel fittings.

On pipes too large in diameter to use forged steel fittings, how are joints (other than joining one straight pipe to another) made?
Saddles, Tees, Ys, and other similar joints are fabricated by cutting and joining the pipe material rather than using fittings of a separate material. These fabrications are most easily done in the shop and then added to the surrounding straight sections, Figure 3–29. If a saddle must be added to an existing pipe, the saddle can be added in the field.

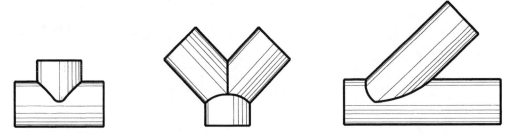

Figure 3–29. Saddle joints on large diameter pipe.

Pipe Flanges

What are welded pipe flanges and why are they used?
Flanges provide a means to rapidly connect pipe either to another length of pipe or to equipment such as valves, filters, meters, tanks, flame arrestors, pumps, or other process equipment. Once the flanges have been put on the pipe, usually by welding, bolts hold one flange to another. When complete, no further welding is required to access or change out flanged equipment. Flanges are often used between straight sections of pipe when clogging is likely and the pipe must be taken apart to clear the clog.

What are the most common types of pipe flange designs and their characteristics?
They are the:
- *Slip-on flange*—This flange slips over the pipe and is then welded both on the inside and outside of the flange.
- *Lap joint flange*—This flange is identical to the slip on, except that the bore is larger. Here is how it is used: The lap joint flange is slipped over the pipe to be flanged, pushed back out of the way, then a stub end is welded to the pipe. Finally the lap joint flange is used to pull the welded stub end against another flange. If exotic and expensive metals such as stainless steel, Hastelloy®, or Monel® are needed in the areas of fluid contact, only the stub flange need be of this expensive material. The lap joint material can be of less expensive material as it does not contact the fluid.

- *Socket weld flange*—There is a stop (or step) on the gasket end of this flange that forms a socket. The pipe slips in up to the end of the socket and is welded on the outside with a fillet weld. These are usually used on high-pressure systems pipe under 4 inches (100 mm) diameter.
- *Threaded or screwed flange*—Usually used where welding cannot be performed.
- *Welding neck flange*—These are sometimes called *high hub* flanges. It is a particularly strong design and transfers forces on the flange to the pipe because the flange-to-pipe joint can be a double-V full-penetration weld. Its high cost limits its use to where strength is essential.
- *Blind or blank flange*—This flange has no opening and is used to close off the ends of a piping system or the flanged outlets of a piece of processing equipment.

See Figure 3–30.

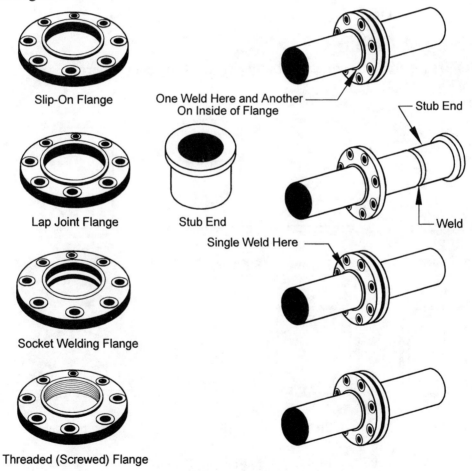

Figure 3–30. Common flange designs. (continued on next page)

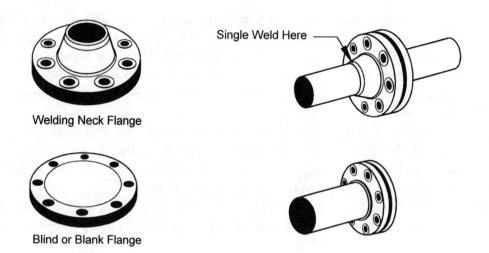

Figure 3–30. Common flange designs. (continued)

Flanged Fittings

Besides pipe-end flanges, what other flange fittings are there?
See Figure 3–31.

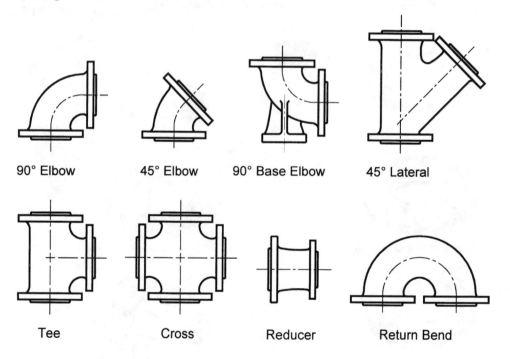

Figure 3–31. Flanged fittings.

Tools for Pipefitting

You are cutting a large diameter pipe to length and need to mark your cut line. How can you be sure the cut line is square with the axis of the pipe?

Use a *wrap-around,* which is a length of thin, flexible material—vinyl, fiber, or cardboard that is wrapped around the pipe 1½ times as shown in Figure 3–32. Adjust the wrap-around so that the second, overlapping layer lies squarely over the first layer, then hold the wrap-around tightly with one hand and mark the cut line with the other. A typical wrap-around is 1/16 inch thick by 2¼ inches wide by 48 inches long (1.6 mm thick, 6 cm wide, by 120 cm long).

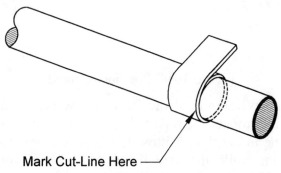

Figure 3–32. Using a wrap-around to get a square cut line.

You need to join two lengths of pipe and want to align them accurately, but you do not have a commercial pipe welding fixture. How is this done?

Tack two lengths of angle iron to form a double V-base as in Figure 3–33. Many applications will require a longer welding fixture than the one shown. Align the pipes for tack welds, make the root pass, then the cover pass.

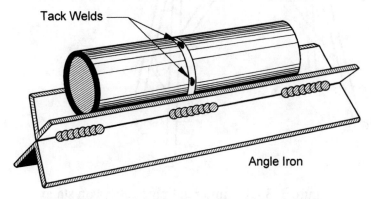

Figure 3–33. Using angle iron to align pipe.

What devices do weldors commonly use to make working on pipe easier?
Pipe jacks, also called pipe stands, support piping at a convenient working height for weldors, so they can weld in a comfortable standing position. They also hold the sections to be joined in the proper position in relation to each other to make tack welds. Weldors may use as few as two and as many as six or seven jacks at one time, Figure 3–34.

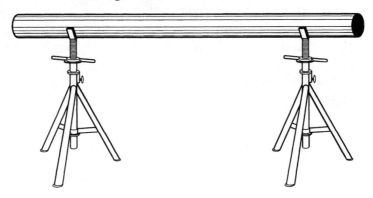

Figure 3–34. Pipe jacks in use.

Pipe rollers permit weldors to always weld in the flat position. The weldor welds while a helper rotates the pipe to keep the weld area on top of the pipe, so welding is done in the flat position. Using four rollers together keeps the pipes in alignment while they are being welded. On very large pipe or castings, a motor driven positioner takes the place of rollers. See Figure 3–35.

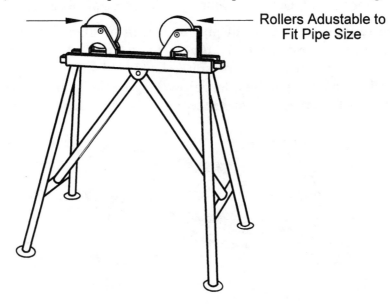

Figure 3–35. Commercial pipe rollers on stand.

On larger diameter pipes, separate fittings are often not used. Connections between pipes like Y's and T's, Figure 3–29, are fabricated from pipe. How does the weldor know what shape cuts to make on pipe so these pipes will fit properly when these joint are fabricated in the field?

The shape—and pattern—for making the intersection of two or more pipes can be determined using a drafting technique called descriptive geometry. From a practical standpoint, there are commercial template kits available made of cardboard or fiber. The weldor selects the proper template, positions it on the pipe, and traces the outline of the template on the pipe for cutting lines. These templates do not cover all possible intersection angles and sizes, but are adequate for most work.

Are there other ways to develop these cutting lines?

Yes, there are several PC programs to do the job. The user inputs the data on the intersecting pipes and surfaces, and then the program generates a printed pattern and a numerical table to mark the cutting lines. These programs are fast and easy to use and produce cutting patterns for any combination of pipe sizes and intersection angles.

Here are the types of problems this software can easily solve:

- Pattern for the intersection of two pipes at any angle, Figure 36–3(a).
- Pattern for the intersection of a pipe with a flat surface; this is useful in structural applications as well as pipefitting Figure 36–3(b).
- Overall pipe length (OAL) and cutting pattern for each pipe end for adding a new pipe between two existing parallel pipes, Figure 36–3(c).
- OAL and cutting pattern for each end for running a pipe between two existing pipes at any angle, Figure 3–36(d). This problem is common in structural pipe fabrication.

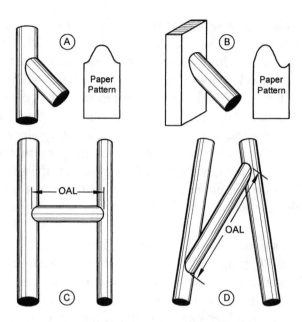

Figure 3–36. Types of pipe intersection problems for which PC-based software generates cutting patterns and related calculations.

What tool can quickly locate and mark any angular point around a pipe?
A Curv-O-Mark® contour tool, shown in Figure 3–37, locates any angle on the outside of the pipe. Just set the adjustable scale to the angle desired, place the tool on the pipe to the "level" position of the bubble, and strike the built-in center punch to mark the point. Figure 3–37, right, shows the contour tool locating the exact top, or 0° position, the 90° position, and the 60° position.

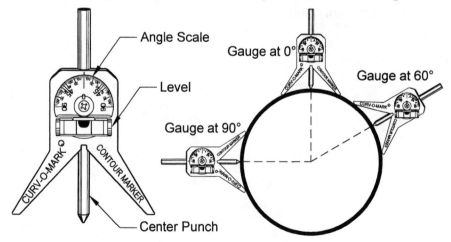

Figure 3–37. Curv-O-Mark® contour tool for locating angular positions around a pipe.

Section II - Tubing

Joining Methods for Tubing

What are the most common methods of joining lengths of tubing?

Copper tubing is most often soldered with an air-acetylene, air-propane, or air-MAPP® gas torch using tin-lead or tin-antimony solder. Tubing for potable water *must* use tin-antimony or other lead-free alloy solder. Building and fire codes require that copper tubing carrying medical gasses be brazed. The same fittings as the soldered copper tubing are used. Non-welded connections such as flared and compression connections are also used. See Section III.

Stainless steel is usually joined by GTAW. While manual GTAW is an effective and reliable joining method, an automated GTAW process called *orbital welding* is gaining popularity. It requires less operator hand skill and produces a printed record of the welding parameters. Because tubing has thin walls, *autogenous* butt welds are used. No filler material is added to the joint, as sections are merely fused together. PAW is also used sometimes.

Aluminum can be brazed or joined with compression or flanged fittings. Steel and steel alloy tubing, usually used in high-performance structures such as racing cars and light aircraft rather than fluid handling, may be oxyacetylene, GMAW, or GTAW welded. Because these applications are structural and load-bearing, careful fitting of the tubing is essential. Flared or compression fittings are often used to join steel and alloy tubing in fluid handling applications.

Copper Tubing Types

When specifying copper tubing, what are the major parameters beside diameter?

The major parameters are:
- Copper tubing is available in *drawn* or *annealed* condition. Drawn copper tubing cannot be bent without having its sidewalls collapse. It comes only in straight lengths of 12, 18, or 20 feet and is ideal for straight runs where appearance is important. Annealed tubing comes in both coils and straight lengths. While annealed tubing can be bent without tools in its smaller sizes, using a tubing bender prevents the tubing from flattening on bends. Because it can be formed to bend or fit around obstacles along its path, many fittings (and copper soldering joints) can be eliminated by using a *single* piece of annealed tubing.

- There are three common wall thicknesses: Type K (heaviest), Type L (standard), and Type M (lightest). All three are used in domestic water service and distribution. Customer budget, preference, and local codes govern this choice. In each case, the actual outside diameter is 1/8 in. (3.2 mm) larger than the nominal or standard size.
- Table 3–6 shows the tube types, color codes marked on the tube lengths, the applicable ASTM Standard, and the commercially available lengths. Other sizes and lengths are available on special order.

Tube type	Color code	Standard	Application	Commercially available lengths		
				Nominal or standard sizes	Drawn	Annealed
Type K	Green	ASTM B 88[3]	- Domestic water service and distribution - Fire protection - Solar - Fuel/fuel oil - HVAC - Snow melting	Straight lengths:		
				1/4 inch to 8 inch	20 ft.	20 ft.
				10 inch	18 ft.	18 ft.
				12 inch	12 ft.	12 ft.
				Coils:		
				1/4 inch to 1 inch	–	60 ft.
					–	100 ft.
				1 1/4 inch and 1 1/2 inch	–	60 ft.
				2 inch	–	40 ft.
					–	45 ft.
Type L	Blue	ASTM B 88	- Domestic water service and distribution - Fire protection - Solar - Fuel/fuel oil - Natural Gas - Liquefied petroleum (LP) gas - HVAC - Snow melting	Straight lengths:		
				1/4 inch to 8 inch	20 ft.	20 ft.
				12 inch	18 ft.	18 ft.
				Coils:		
				1/4 to 1 inch	–	60 ft.
					–	100 ft.
				1¼ inch to 1½ inch	–	60 ft.
				2 inch	–	40 ft.
					–	45 ft.

Table 3–6. Copper tube types, applications, and lengths.
(continued on next page)

Tube type	Color code	Standard	Application	Commercially available lengths		
				Nominal or standard sizes	Drawn	Annealed
Type M	Red	ASTM B 88	- Domestic water service and distribution - Fire protection - Solar - Fuel/fuel oil - HVAC - Snow melting	Straight lengths:		
				¼ inch to 12 inch	20 ft.	N/A
DMV	Yellow	ASTM B 306	- Drain, waste, vent - Solar - HVAC	Straight lengths:		
				1¼ inch to 8 inch	20 ft.	N/A
ACR	Blue	ASTM B 280	- Air conditioning, refrigeration - Natural gas - Liquefied petroleum (LP) gas	Straight lengths:		
				3/8 inch to 4-1/8 inch	20 ft.	4
				Coils:		
				1/8 inch to 1-5/8 inch	–	50
OXY, MED OXY/MED OXY/ACR ACR/MED	(K) Green (L) Blue	ASTM B 819	- Medical gas	Straight lengths:		
				¼ inch to 8 inch	20 ft.	N/A
Type G	Yellow	ASTM B 837	- Natural gas - Liquid petroleum (LP) gas	Straight lengths:		
				3/8 inch to 1-1/8 inch	12 ft.	12 ft.
					20 ft.	20 ft.
				Coils:		
				3/8 inch to 7/8 inch	60 ft.	
					100 ft.	
					–	100 ft.

Table 3–6. Copper tube types, applications, and lengths. (continued)

Soldered or Brazed Tubing Joints

What is the basic design of soldered or brazed tubing joints?

These joints most often use wrought copper or cast copper-alloy fittings which slip over the end of the tubing and form a cylindrical bond on the tubing when solder is applied, Figure 3–38. A wide variety of these fittings are available with straight splices, Tees, Ls, crosses, caps, 45° angles, and adapters to fit between different pipe sizes. Fittings with solder (or braze) joints on one end and pipe threads on the other are available to connect copper lines to steel and plastic pipe. Also, many valves and other devices are connected through solder or braze fittings.

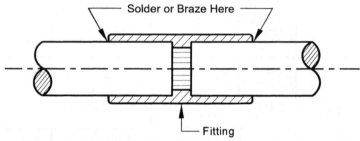

Figure 3–38. Copper tubing fitting for soldering or brazing.

Because copper is so malleable, a hand tool may be used to produce an integral copper fitting on the end of a length of tubing by expanding it into a socket with the dimensions of a fitting. This saves a fitting and one soldered joint, Figure 3–39.

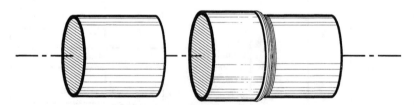

Figure 3–39. Integral expanded socket in tubing.

What determines the choice between soldering or brazing on copper pipe joints?

The choice usually depends on the operating conditions of the system and the requirements of local construction codes. Solder joints are generally used where the service temperature does not exceed 250°F (121°C). Brazed joints can be used where greater joint strength is required or where system temperatures are as high as 350°F (177°C). Note that although brazed joints offer higher joint strength in general, the annealing of the tube and fitting

WELDING FABRICATION & REPAIR

which results from the higher heat (from the oxyacetylene torch) used in the brazing process can cause the rated pressure of the system to be less than that of a soldered joint.

Holding copper tubing (or other thin-walled tubing) in a vise when cutting it with a hacksaw or portable band saw distorts the tubing and the vise jaws damage the smooth sides. How do you avoid this?

First insert a tightly-fitting wooden dowel inside the tubing end to prevent it from collapsing in the vise. Then place *soft jaws* over the steel jaws of the vise to prevent marring the tubing. Soft jaws are usually made of aluminum, copper, lead, or plastic. See Figures 3–40 and 3–41.

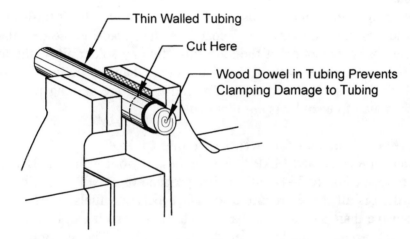

Figure 3–40. Using a dowel when clamping thin-walled tubing.

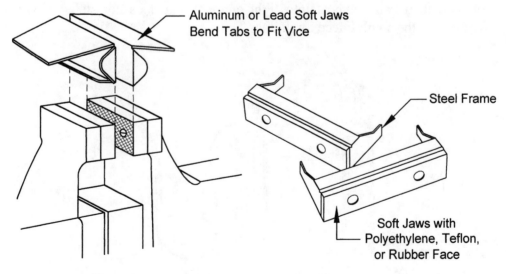

Figure 3–41. Soft jaws prevent damage to the work.

Soldering Copper Tubing

What solders may be used on copper tubing carrying potable water?

By U.S. Federal law, only lead-free solders, pipe, and fittings may be sold or used in drinking water systems. These solders are commonly 95-5 tin-antimony, but may include a variety of other alloys of about 95% tin combined with copper, nickel, silver, antimony, bismuth, or other alloying elements. These lead-free solders can cost 4 to 8 times as much as the tin-lead alloys they replace.

What solders may be used for copper carrying *non-drinkable* water or other fluids?

Lead containing solders may be used. Most common are 50-50 tin-lead and 60-40 tin-lead alloys, because they cost less than lead-free solder alloys. However, with the exception of the 95-5 tin-antimony solder alloy, the lead-free solders have a wider pasty temperature range than the traditional 50-50 and 60-40 tin-lead solders, are easier to handle, and are gaining popularity even where a lead-free solder is *not* required.

What torches are suitable for sweating copper tubing?

Air-acetylene, propane, and MAPP® gas torches all work well. The Bernz-O-matic® torches in Figure 3–42 burn either propane or MAPP gas. Changing the gas orifice is all that is required to switch between fuels. MAPP gas is more expensive than propane and has a hotter flame, so it can heat the pipe and fitting faster and handle larger diameter tubing. The version with a separate regulator, 48-inch (1.2 m) hose, and hand piece is especially convenient. It can put heat into tight quarters and does not stall out when inverted, as the single-piece torch does.

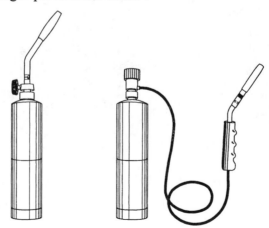

Figure 3–42. Torch-on-cylinder model (left) and hose-based model (right).

WELDING FABRICATION & REPAIR

Are there other heat sources for soldering copper tubing besides torches?
Yes. Where open flames or compressed gas cylinders are a hazard, electrical resistance soldering works well. A step-down transformer supplies a high current to a pliers-like hand piece through a pair of cables. This hand piece clamps carbon blocks against opposite sides of the fitting. High current from the transformer flowing through the copper fitting's resistance generates most of the heat (the rest comes from the resistance of the carbon blocks), Figure 3–43. The process is fast, clean, and flameless. It is a complete replacement for torch, hoses, and cylinders. Solder is applied as with a torch.

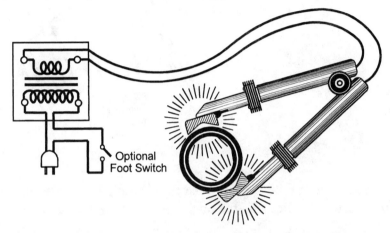

Figure 3–43. Electrical resistance soldering of copper tubing.

What steps are needed to prepare copper tubing for soldering?
- Measure the tubing length for cutting. The tubing must be long enough to reach to the cup (bottom) of its fitting, but not so long as to cause stress in the completed piping.
- Cut the tubing, preferably with a disc-type tubing cutter to insure square ends. A hacksaw, an abrasive wheel, or a portable or stationary band saw may also be used.
- Remove burrs from inside the tubing ends with a round file, half-round file, or reamer. Many disc-type tubing cutters carry a triangular blade for inside reaming. Remove outside burrs with a file because these burrs prevent proper seating of the tubing into the fitting cup. A properly reamed piece of tube provides an undisturbed surface through the entire fitting for smooth, laminar flow, and minimum pressure drop.
- Use emery cloth, nylon abrasive pads like 3M™ Scotch-Brite®, or male and female stainless steel brushes sized to the tubing as in Figure 3–44, to remove the dirt and oxide from the end of the tubing and the inside mating surfaces of the fitting. An electric drill can drive the male stainless steel internal brush and save a lot of elbow grease on a large job. The goal is to

get down to fresh, shiny metal. Clean the outside of the tubing from the end of the tubing to about 3/8 inch (1 cm) beyond where the tubing enters the fitting. The capillary space between the tubing and its fitting is about 0.004 inches (0.1 mm). Filler metal fills this gap. Removing too much metal from either the tubing or the fitting will prevent proper flow of solder around the tubing by capillary forces and weaken the joint. Do not touch the cleaned area with bare hands or oily gloves. Skin oils, lubricating oils, and grease impair the adherence of filler metal.

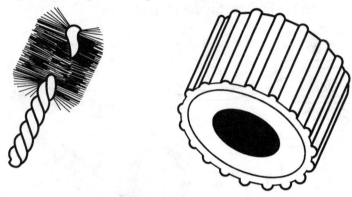

Figure 3–44. Male stainless steel brush for cleaning inside fittings (left) and female brush for cleaning the outside ends of copper tubing (right).

- Select the flux to match the solder, since a mismatch may cause problems. Use a flux meeting ASTM B 813 requirements.
- Using a brush, apply a thin coating of flux to the ends of the tubing sections and the fitting's interior mating surfaces. A disposable acid brush works well. Keep the flux off your hands, since chemicals in the flux can be harmful if carried to the eyes, mouth, or open cuts. The soldering flux will dissolve and remove traces of oxide from the cleaned surfaces to be joined, protect the cleaned surfaces from re-oxidation during heating, and promote wetting of the surfaces by the solder metal.
- Insert the tubing end into the fitting with a twisting motion. Make sure the tubing is properly seated in the fitting cup. With a cotton rag, wipe excess paste flux off the tubing from the exterior of the joint. Caution: Do not leave a cleaned and fluxed joint unsoldered overnight. If you must stop work before soldering, disassemble the joint(s) and remove all the flux. On starting work, the next day, perform the cleaning and fluxing process from the beginning.
- Most of the time there are two or three tubing lengths going in a fitting. Insert all the tubing lengths into the fitting so they can all be soldered in one operation. This will save work time and torch fuel. Make sure the tubing lengths and fittings are supported or braced so that all tubing

remains at the bottom of its respective fitting cup. You are now ready to solder the fitting.

How is the fitting and tubing heated?
- Begin by heating the copper tubing outside and just beyond the fitting, Figure 3–45. This brings the initial heat into the fitting cup and provides even heat to the joint area. If the joint is vertical be sure to heat the tubing around its entire circumference. If the joint is horizontal, heat the bottom and sides of the tubing. Do not heat the top of the tubing to avoid burning the flux. The natural tendency of heat to rise preheats the top of the assembly and brings sections of tubing up to temperature. The larger the joint, the more heat and time will be needed to accomplish preheating. Experience will determine the amount of preheat time needed. Large-diameter tubing is best soldered with two torches or a multi-orifice torch.

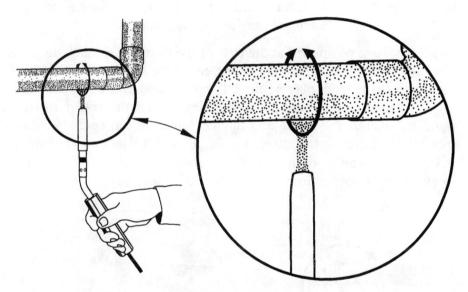

Figure 3–45. Preheat the tubing around the bottom and sides about one cup length away from the fitting.

- Now move the flame onto the base of the fitting cup and sweep the flame back and forth between the base of the fitting cup and along the tubing sides to a distance equal to the depth of the fitting cup, Figure 3–46. Preheat the circumference of the assembly as described above. Again, do not heat the top during this preheating to prevent burning the flux.

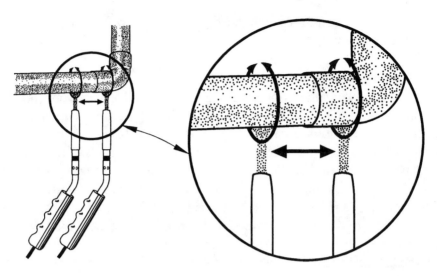

Figure 3–46. Preheat the fitting starting from the base of the cup and sweep out to the previous preheat zone.

- With the flame positioned at the base of the fitting cup, touch the solder to the joint at the point where the tubing enters the fitting, Figure 3–47. Keep the torch flame from playing directly on the solder. If the solder does not melt, remove it, and continue preheating. When the solder melts when touching against the fitting and tubing, soldering can begin. Caution: Do not overheat the joint or direct the flame onto the face of the fitting cup. Overheating could burn the flux, destroying its effectiveness, and preventing the solder from entering the fitting properly.

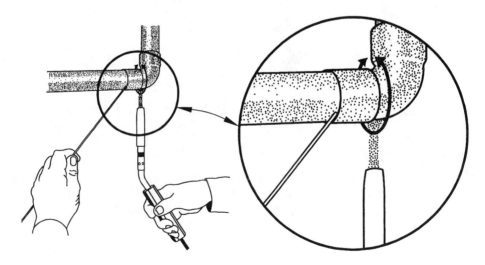

Figure 3–47. Determining when preheating ends and soldering begins.

Welding Fabrication & Repair

How is the actual soldering performed?

For joints in the horizontal position:

- Start applying the solder slightly off-center at the bottom of the joint, as in Figure 3–48. When the solder begins to melt, push the solder into the joint while moving the torch upward along the base of the fitting and slightly ahead of the point of application of the solder. When the bottom portion of the fitting fills with solder, a drip of solder will appear at the bottom of the fitting. Continue to feed solder into the joint while moving the solder up and around to its top or 12 o'clock position, and at the same time move the flame so it slightly leads the solder up to the top of the joint. In general, solder dripping off the fitting bottom is not coming from the bottom of the joint. As the joint fills, excess solder runs down the face of the joint and drips off the bottom. This is a good indication of using the proper amount of solder and a full joint. This will allow the solder in the *lower* portion of the joint to become pasty, form a dam and support the solder applied above it, Figure 3–49 (a) and (b).

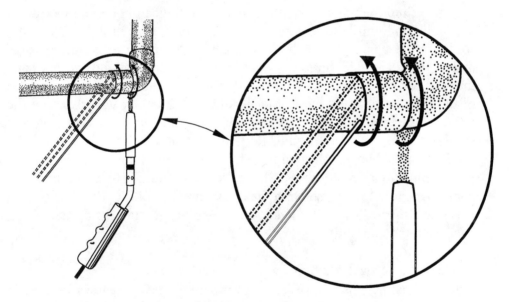

Figure 3–48. Soldering up the side of the fitting.

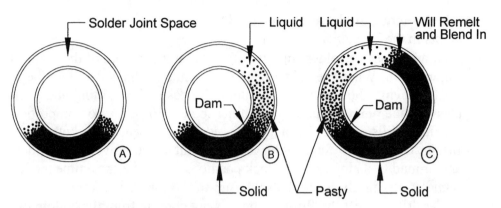

Figure 3–49. Sectional view of a solder fitting showing how cooler solder in the base of fitting forms a dam to hold pasty solder in place until it cools.

- Move torch heat and solder to the other side of the fitting and overlapping slightly, proceed up the uncompleted side to the top and over, again overlapping slightly, Figure 3–49 (c). While soldering, small solder drops may appear behind the point of solder application, indicating the joint is full to that point and will take no more solder. Because these drops indicate that solder in the upper part of the joint is above the pasty temperature range and in the full liquid state, we want to cool the joint slightly to get back into the pasty temperature range. To do this, relocate the flame to the base of the fitting and apply more solder. Adding this solder will also cool the joint. Throughout this process you are using all three physical states of solder: solid, pasty, and liquid. Remember that you want to control the state of the solder: cooler solidified solder at the bottom of the fitting, pasty transitional state above the solid, and liquid solder at point of application to allow the solder to be drawn into the fitting. Beginners often apply too much heat and cannot get the solder to stay inside the fitting because it is too liquid.

For joints in the vertical position:
- After preheating, make a series of overlapping circular passes feeding in solder and starting wherever is convenient. Stop applying solder when you can see the joint is filled and it will take no more solder. Vertical joints are the easiest to start on.
- Since vertical joints are the easiest to make, it is often wise to plan the job so the maximum number of joints are vertical. Plumbers often preassemble sections of tubing and fittings on a bench or in a vise where it is more convenient and comfortable to work.

For fittings with horizontal or cup-inverted joint:
- Begin soldering the bottom-most joint first since the heat applied to it will preheat the joints above it. If you solder the upper joints first, you will melt the solder out of them when you begin soldering the lower joints as the convected heat rises. By soldering the lowest joint first, you not only preheat the joint(s) above it, but you have total control of the heat input to this joint. If you overheat this joint, the solder will be too thin (runny) to be held in place by capillary attraction and it will run out. Just the right amount of heat will let the solder be drawn up into the joint. After the lowest joint is done, the side or top joint will then be ready to solder. Apply a little more heat to it and run the solder around the ring where the tubing meets the fitting on the side, then the top joint.
- In all cases, trying to improve the integrity of a soldered joint by loading additional solder where the tubing enters the fitting is pointless. The strength of the joint is the solder bond between the inside of the fitting and the outside of the tubing inside the fitting. If the joint was clean, properly fluxed and heated when the solder was applied, it will hold.

What steps are needed to complete the joint?
- First, allow the assembly to air cool without disturbing it. Do not spray it with water or dip the fitting in water as cracking may result.
- Then, once the joint has reached room temperature, take a clean, cotton rag dipped in water or alcohol and wipe the flux residue from the joint. Wiping the joint while the solder is still molten can lead to joint cracking and is not recommended. Excess solder should be removed with the torch while the joint is still hot. Figure 3–50 shows what a perfect male joint should look like.

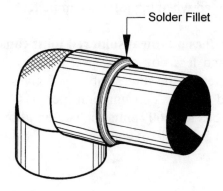

Figure 3–50. A perfect male joint.

- Finally, test the completed joint for leaks, and flush the tubing lines to remove the remaining flux.

What steps must be taken to solder tubing to a valve with solder cups?
Soldering must be done in the open position. Valve stems are removed to prevent exposing the packing and washers to torch heat. Most ball valves can be soldered without disassembly as they have Teflon® PTFE seals which withstand soldering temperatures for a short time.

What steps must be taken when soldering tubing lines that have been in service and contain water?
Soldering must be done in tubing and fittings that are free of water. Some methods to achieve a dry pipe when the line has been in service are:
- Open taps above and below the joint to help the tubing drain.
- Remove pipe straps on each side of the joint to permit bending the tubing down to let water run out.
- On vertical lines containing water, use a short length of 1/4-inch hose to siphon out the water so the water level inside the line is at least 8 inches below the joint.
- Using paper towels and a stick, dowel, or long screwdriver, remove all moisture and water drops within 8 inches (200 mm) of the soldering.
- Whenever you see steam rising from a fitting after sweating, there is a very good chance that water in the lines has spoiled the joint. Take it apart, fully drain the lines, and begin again.
- Valves on the lines to be soldered may not shut off completely, or it may be impossible to drain the line enough to prevent all water from entering the soldering area. To keep this water away, stuff fresh white bread (cut off crusts as they dissolve slowly) into the line(s) to form a plug and push it 8 inches (200 mm) back into the line away from the fitting. Then assemble the copper tubing to the fitting and complete the solder joint. Flush out the bread when the joint is complete.

How much solder does a typical soldered joint consume?
For joints 1 inch and less you can expect to use *roughly* a length of solder equal to the diameter of the joint, Table 3–7 gives the exact length. It also provides estimated solder consumption per 100 joints for job costing. Estimates in the table for 100 joints include an allowance for wastage. Flux should be estimated at 2 oz/lb of solder.

Nominal or standard size inches	O.D. of tube, inches	Cup depth of fitting, inches	Joint Clearance, Inches[1]		Wt. in lbs. at 0.010 inches clearance per 100 joints
			0.005	0.010	
1/4	0.375	0.310	0.149	0.298	0.097
3/8	0.500	0.380	0.243	0.486	0.159
1/2	0.625	0.500	0.400	0.800	0.261
5/8	0.750	0.620	0.595	1.191	0.389
3/4	0.875	0.750	0.840	1.680	0.548
1	1.125	0.910	1.311	2.621	0.856
1 1/4	1.375	0.970	1.707	3.415	1.115
1 1/2	1.625	1.090	2.268	4.535	1.480
2	2.125	1.340	3.645	7.291	2.380
2 1/2	2.625	1.470	4.940	9.880	3.225
3	3.125	1.660	6.641	13.282	4.335
3 1/2	3.625	1.910	8.864	17.728	5.786
4	4.125	2.160	11.41	22.813	7.446
5	5.125	2.660	17.45	34.905	11.392
6	6.125	3.090	24.23	48.459	15.815
8	8.125	3.970	41.29	82.589	26.955
10	10.125	4.000	51.85	103.696	33.845
12	12.125	4.500	69.85	139.701	45.596
			Average Actual Consumption of Solder in Inches	Allowance for Estimating of Solder in Inches	

Note 1: Using 1/8-inch diameter (No. 9) wire solder (1 inch = 0.01227 cubic inches)

Table 3–7. Solder consumption.

Brazing Tubing Joints

What are the two classes of brazing filler metals and what is the difference between them?

The two classes of brazing filler metals are BCuP and BAg. The *BCuP* alloy series contains phosphorus, while the *BAg* series has a high silver content. The two classes differ in their melting, fluxing, and flowing characteristics. The BCuP series of brazing alloys are suited to general piping applications where copper is being joined to a similar alloy such as brass or bronze. The members within this series are selected to match the fit, or joint tolerance. The BAg alloys are for joining dissimilar metals such as steel or stainless steel. This alloy series is more expensive than the BCuP series.

How do brazing fluxes differ from soldering fluxes?
Brazing fluxes are formulated to work at higher temperatures than soldering fluxes. Brazing fluxes are water-based, while soldering fluxes are often petroleum-based.

How does the brazing of copper tubing differ from its soldering?
- In brazing copper tubing, an oxyfuel torch is usually used for heat instead of an air-propane, air-MAPP® gas or air-acetylene torch.
- Brazing filler metals melt in the 1100 to 1700°F (593 to 927°C) temperature range; solder melts in the 360 to 650°F (182 to 343°C) range.
- Brazing fluxes should meet AWS Standard A5.31, Type FB3-3 or FB-3C.
- The process of applying heat and filler metal is quite similar to that used for solder.

Troubleshooting Soldered or Brazed Joints

What are the most common causes of failures in soldered or brazed joints?
Although soldering and brazing operations are inherently simple, the omission or misapplication of any single part of the process may mean the difference between a good joint and a failure. Faulty joints usually result from one or more of the following factors:
- Improper joint preparation prior to soldering.
- Lack of proper fitting and tubing support during soldering or brazing may cause wiggling of the joint while the filler metal is molten. Non-concentric joint alignment may lead to gaps in filler metal.
- Improper heat control and heat distribution through the entire joining process.
- Improper application of solder or brazing filler metal to the joint.
- Inadequate filler metal applied to the joint.
- Sudden shock cooling and/or wiping the molten filler metal following soldering or brazing.
- Pretinning of joint prior to assembly and soldering.

You must solder copper tubing fittings in place in a wall. When the torch is used to apply heat to the fitting, the surrounding materials—the 2×4"s and the drywall—catch fire. How do you prevent this?
Use a commercial woven glass heat shield to keep the torch flame off the

WELDING FABRICATION & REPAIR

building materials, Figure 3–51, or use a piece of heavy steel or aluminum to do the same thing. Old timers used pieces of asbestos and these worked even better. There are also commercial products in paste and spray form that help control heat from soldering and brazing operations. Examples of these products are *Block-it Heat Absorbing Paste* and *Cool Gel Heat Barrier Spray* from LA-CO®/Marcal.

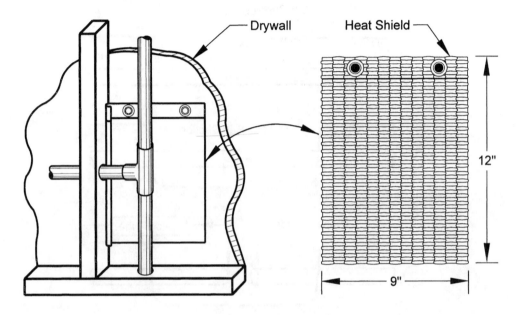

Figure 3–51. Using a heat shield when working in close quarters.

An underground copper tubing water supply line has developed a leak as in Figure 3–52 (a). You have located the section of the pipe to replace and cut it out, Figure 3–52 (b). The problem now is how to install a new section of line and get it seated properly when each end of the copper line is firmly embedded in the soil and cannot be moved *horizontally*. How to proceed?

- Purchase a piece of special diameter patch tubing, also called a repair coupling. It is made to just slip *over* the *outside* of the existing tubing; it is sold in short lengths.
- Cut a piece of this patch tubing about 2 inches (50 mm) longer than the gap in the line.
- Prepare the outsides of the ends of the line and the insides of the ends of the patch tube by cleaning and fluxing, Figure 3–52 (c).
- Expose enough of one end of the existing copper tubing so it can be pulled up and the section of patch tubing can be slipped on it, Figure 3–52 (d).
- Center the patch tubing over the break.

- Solder the patch in place and the repair is complete, Figure 3–52 (e). This is a permanent repair.

Note: A temporary, emergency repair may be made by cleaning the area of the crack and brazing the joint area with a silver-containing filler metal.

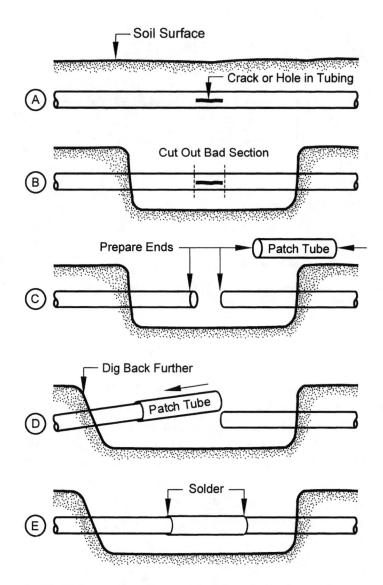

Figure 3–52. Using a slip-over patch tube when pipe ends cannot be separated.

Copper Tubing Installation Issues

How should copper tubing be installed underground?
Plumbing codes and good piping practice require that all excavations be completely backfilled as soon after inspection as practical. Trenches should first be backfilled with 12 inches of tamped, clean earth that should not contain stones, cinders or other materials that would damage the tube or cause corrosion. Equipment such as bulldozers and graders may be used to complete backfilling. Suitable precautions should be taken to ensure permanent stability for tube laid in fresh ground fill. Do not fill the tubing trench with construction waste materials.

At what intervals should copper tubing be supported?
Drawn temper tubing, because of its rigidity, is preferred for exposed piping. Unless otherwise stated in plumbing codes, drawn temper tube requires support for horizontal lines at about 8-foot intervals for sizes of 1 inch and smaller, and at about 10-foot intervals for larger sizes. Vertical lines are usually supported at every story or at about 10-foot intervals, except for long lines, where there are the usual provisions for expansion and contraction, anchors may be several stories apart, provided there are sleeves or similar devices at all intermediate floors to restrain lateral movement.

Annealed temper tube in coils permits long runs without intermediate joints. Vertical lines of annealed temper tube should be supported at least every 10 feet. Horizontal lines should be supported at least every 8 feet.

Can copper tubing withstand freezing?
Annealed temper copper tubing can withstand the expansion of freezing water several times before bursting. Under test, water filling a 1/2-inch soft tube has been frozen as many as six times, and a 2-inch size, eleven times. This is a vital safety factor favoring soft tube for underground water services. However, it does not mean that copper water tube lines should be subjected to freezing; steps should be taken to prevent freezing.

What causes water hammer and how can it be prevented?
Water hammer is the term used to describe the destructive forces, pounding noises, and vibrations that develop in a water system when the flowing liquid is stopped abruptly by a closing valve. When water hammer occurs, a high-pressure shock wave reverberates within the piping system until the energy has been spent in frictional losses. The noise of such excessive pressure surges may be prevented by adding a capped air chamber or surge arresting device to the system.

Arresting devices are available commercially to provide permanent protection against shock from water hammer. They are designed so the water in the system will not contact the air cushion in the arrester and, once installed, they require no further maintenance.

On single-fixture branch lines, the arrester should be placed immediately upstream from the fixture valve. On multiple-fixture branch lines, the preferred location for the arrester is on the branch line supplying the fixture group between the last two fixture supply pipes.

How much does copper tubing expand when heated and what provisions should be made for this expansion?
A 100-foot length of copper tubing experiencing a 100°F temperature rise will increase its length by 1.13 inches. In metric terms, a 50-meter length of copper tubing seeing a 50°C temperature rise will increase its length by 42.5 mm. Including a loop, U-bend, or offset in a long length of tubing provides a way for the expansion to take place without damaging the tubing, Figure 3–53.

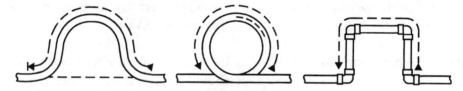

Figure 3–53. Adding a U-bend (left), coiled loop (center) or offset and return (right) provides room for expansion with temperature changes.

Orbital Welding
What is orbital welding and how does it work?
Orbital welding is an automated GTAW process for pipe and tubing, principally steel, stainless steel, and exotic alloys. The typical system welds 1/8- through 6-inch (3 through 150 mm) diameter material. The main applications are high-pressure tubing in aerospace and power station applications, high-purity tubing in semiconductor manufacturing, and sanitary tubing in food, beverage, diary, and pharmaceutical industries. While many older analog orbital welding systems are still in use, the latest designs are digital and use a microprocessor-controlled constant-current power supply and a welding head in which the welding occurs. The operator communicates with the system through a keyboard and LCD display. System software suggests a welding cycle based on the size, joint type, and work metal, but the weldor can alter and fine tune the welding variables such as weld speed, number of revolutions, start delay, current level, tacks, and

pre/postweld inert gas flow. A cable connects the power supply with the welding head to provide welding current, motor control lines, and inert gas. A clam shell design permits the head to open, fit over the tubing (or pipe) ends, then close to clamp them rigidly in coaxial alignment.

Once the weldor clamps the ends in the welding head and turns on the welder, the process is fully automatic. Inert gas fills the head interior to protect the molten weld metal from the atmosphere. A DC motor drives a tungsten electrode circumferentially around the outside of the tubing (or pipe) joint to make the weld. The microprocessor monitors the progress of the weld to assure the set parameters were accomplished. Some systems have a small printer in the welding power supply to provide a permanent record of the weld parameters. Parameters can also be stored, or transferred to other machines. Figure 3–54 pictures an orbital welding power supply and its welding head, Figure 3–55 diagrams how welding is done inside the welding head itself.

What are the advantages of orbital welding?
- All the advantages of GTAW: high-quality welds without spatter, smoke, slag, or filler metal.
- Produces high-quality welds with smooth interiors that prevent debris from hanging up and accumulating *inside* the pipeline.
- Fully automated with welding parameters monitored and verified during welding, permanently documented, all under microprocessor control.
- Weld quality not dependent on weldor's hand skills.
- Faster than hand GTAW process.
- Process makes welds in tighter confines than human weldor.
- Orbital welding equipment readily moved by one weldor.

What are the drawbacks to orbital welding?
- Requires a larger volume of welds to justify the equipment cost.
- Tube ends must be machined square, no saw cuts permitted.
- High quality, low production rate process.

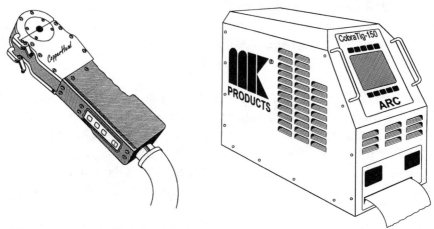

Figure 3–54. Orbital welding head (left) and power supply (right).

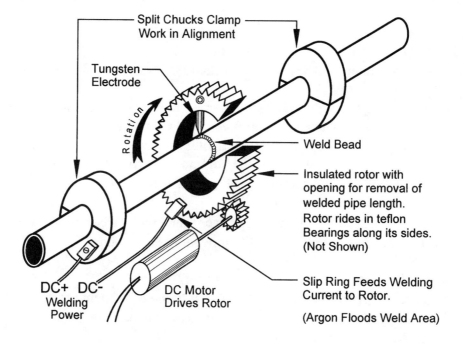

Figure 3–55. Simplified schematic of orbital welding inside the welding head.

Fabricating Structures with Tubing

What is a *saddle* and why is it made?

A saddle (or *fishmouth*) is the shaping of the end of one piece of tubing so it meets and fits tightly against another. We do this to make strong welds. The gap between the two tubing pieces should not exceed the diameter of the GMAW/FCAW/GTAW wire used to make the weld. See Figure 3–56.

WELDING FABRICATION & REPAIR

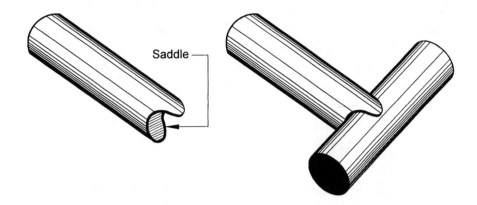

Figure 3–56. Tubing saddle.

What are several common ways to make a tubing saddle?
- Hacksawing and hand filing.
- Beginning the saddle on a bench or pedestal grinder, making the final adjustments with a file.
- A hand-operated nibbling tool is a faster way to make a saddle, although some operator skill is needed.
- Using a saddling hole-saw tool works well, Figure 3–57.

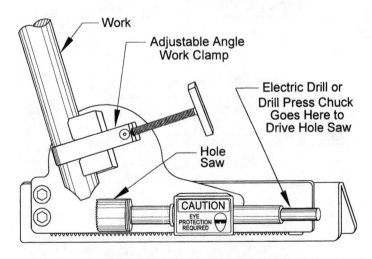

Figure 3–57. Saddle cutter.

Steel Structural Tubing Repairs

What is a good way to repair cracked *structural* tubing that carries no fluid?
- Drill 1/4-inch (6.5 mm) stress relief holes at the ends of the crack.

- Cut or grind away the cracked area between the ¼ inch holes.
- Cut a patch out of a similar-sized piece of tubing and reshape this patch to fit over the cracked tubing area.
- Weld on the patch as shown in Figure 3–58. Use no continuous welds, no end welds and no welds closer to the end of the patch piece than ¼ inch (6.5 mm).

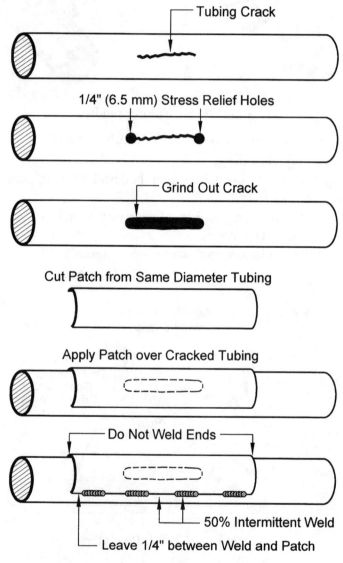

Figure 3–58. Repairing cracked structural tubing.

Aluminum Tubing Repairs

How can punctured aluminum tubing be repaired?

There are several approaches:
- The damaged section can be cut out and a new length spliced in with flared or compression fittings.
- The hole can be covered with a soldered patch. First, file the punctured area so it is flush with the rest of the pipe's surface, then clean the aluminum pipe with emery cloth down to clean metal. Using steel wool, and a propane or MAPP® gas torch, tin the aluminum tubing by heating the tubing, scrubbing the steel wool back and forth over the pipe. You will want to hold the steel wool with a long-nose pliers. After several passes, the aluminum will become tinned or wetted with solder. A 50-50 tin-lead or 60-40 tin-lead alloy works fine. Cut out a patch of brass and form the patch to fit on the pipe. Tin the patch, then heat both the patch and the tubing, and apply the patch over the previously tinned area of the aluminum tubing. See Figure 3–59.

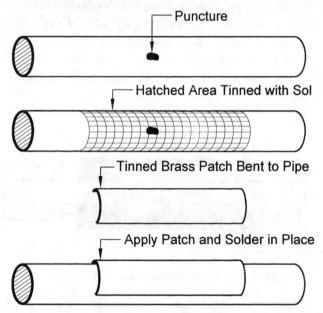

Figure 3–59. Temporary repairs for punctured aluminum tubing with tin-lead solder.

Section III – Non-Welded Pipe Joining Fittings

What are the most common non-welded fittings?

- Flared connections, Figure 3–60, are used for copper, aluminum, or steel tubing. These are widely used for water, oil, fuel, and some hydraulic lines. Copper is restricted to low pressures and must not be used in vehicles where vibration will work-harden and eventually crack it. Steel—either terne (tin) or copper plated—is used for high pressure-hydraulic lines in vehicles and machinery. Flaring tools are inexpensive and easy to use. The fittings are cheap and widely available.

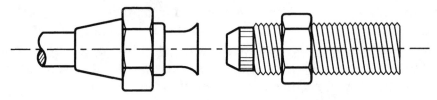

Figure 3–60. Flared tubing connection.

- Compression fittings are often used in plumbing and work well on both copper and brass tubing, Figure 3–61. Other than a disc-type tubing cutter to make a square end, no special tools are needed to prepare the tubing for installation. A brass ferrule slides over the end of the tubing, and secured inside the fitting by screw threads. Because the tubing and brass ferrule are relatively soft, they form a fluid-tight seal when the fitting is pulled up tight. It is not unusual for considerable force to be needed on the fitting to stop a slow leak.

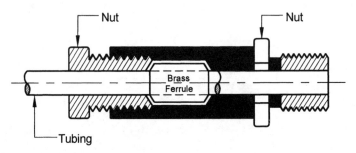

Figure 3–61. Compression tubing connection.

- Victualic® piping systems are frequently used in industrial and commercial sprinkler and process piping. Common steel pipe with special grooves rolled or cut into its ends is joined by a Victualic fitting. The fitting holds the pipes together by gripping grooves on the pipes' ends, and also compresses the rubber gasket inside it to form a fluid seal between the pipes, Figure 3–62. Pipe is usually supplied with grooves, but equipment on site is used to put grooves on cut lengths. Victualic piping systems handle a wide variety of pipe materials including steel, stainless steel, glass-lined steel, cement-lined steel, copper and plastic pipe. Fittings

WELDING FABRICATION & REPAIR

from 1 inch (25 mm) to 30 inches (750 mm) are common and larger sizes are available. The Victualic system has the following advantages:
- Piping installs quickly with few tools.
- No welding is used and no fire hazards are created.
- Fittings work for pressure and vacuum.
- Lines may be readily disassembled if clogged.
- Fittings handle vibration, deflection, misalignment, and permit expansion.
- Fittings available from 150 to 2500 psi (1030 to 17240 kPa)

What important safety step must be observed when working with Victualic pipe fittings?
Always make sure that the system is depressurized *before* attempting to remove a fitting, or serious injury could result.

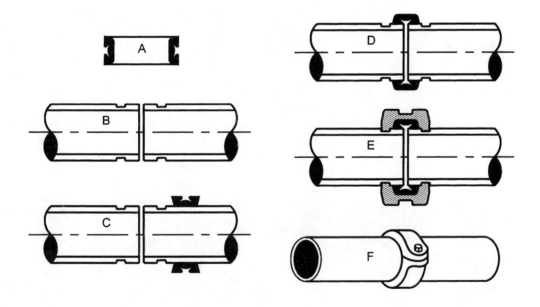

Figure 3–62. Assembling Victualic pipe connection: (a) Gasket turned inside out and lubricated, (b) grooved-end pipe, (c) cross-section of gasket slid onto one pipe inside out, (d) cross-section of gasket centered over joint, right side out, (e) cross-section with outer bolted housing securing gasket and holding pipes together, (f) external view of finished pipe connection.

Chapter 4

Bending & Straightening

*Do you know the difference between education and experience?
Education is what you get when you read the fine print, experience is what
you get when you don't.*
—Pete Seeger

Introduction

Bending operations are used in almost every fabricated metal product and the weldor is often required to perform them as part of his job. Many times the welding cannot be done until a bending step is completed.

This chapter:
- Describes the most common bending equipment for the metal fabricator and details their capabilities and advantages.
- Shows how a mechanical bending machine performs bends in a variety of rolled shapes and tubing and describes the tooling needed. Although this particular machine is small, the same principles hold true in bending operations of any size.
- Explains the theoretical basis of heat bending and straightening operations using an oxyfuel torch and provides enough detail to enable the beginning weldor to try his hand at them.
- Pictures three different types of templates used to check large-scale bends. These tools can easily be shop built and will be helpful in future bending projects.
- Presents some suggestions for handling problems in welded fences and railings.
- Matches bending machines with thicknesses and shapes of metals.

Section I – Bending

Bending Equipment

What are the most common ways to bend metal?
There are many tools specifically designed for bending different types of metals. Here are some of them:
- *Hand benders* are inexpensive and useful for small sheet metal bends. See Figure 4–1.

	Sheet Metal >3/16" Thick	Plate ≤3/16" Thick	Rods & Bars	Shapes: Ts, Ls, Cs, I-Beams	Wire	Pipe & Tube
Hand Bender	●					
Pin Fixtures and Bending Jigs			●▲		●	●▲
Sheet Metal Brakes	●					
Press Brakes	●	●	●			
Roll Brakes		●				
Bending Machines Rotary, Mandrel, Sand Filling	●	●	●▲	●▲	●	● ● ●
Wrinkle Bending			●▲			▲
Rolls	●	●	●▲	●▲		●▲
Flame Bending		▲		▲		▲
Line Heating		▲				
Flame Straightening		▲	▲	▲		▲
Flame Panel Shrinking		▲				

● = No heat   ▲ = Heat used

Table 4–1. Bending methods matched to materials.

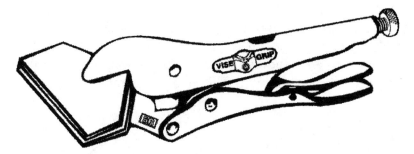

Figure 4–1. Vise Grip® locking sheet metal pliers.

- *Pin fixtures and bending jigs* can be commercially made or shop-built. Inexpensive and easy to use, they work well on wire, rods, and bars. Usually the base is held in a bench-mounted vise, but some are bolted directly to the workbench. Lighter stock can be bent cold, while heavier stock bends more easily when heated.

Changing the pin diameter changes the bend radius. Gradual, long-radius bends are made by making a series of small bends along the workpiece, while S-bends, U-bends, and 90° bends are made in a single operation, Figure 4–2.

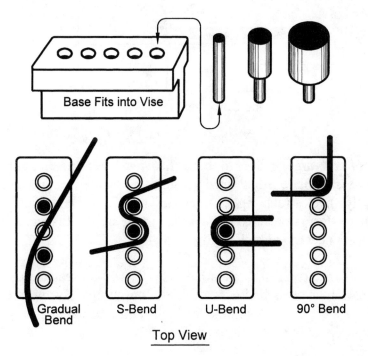

Figure 4–2. Bending jigs work well on wire, rod, and light bar stock.

- *Sheet metal brakes* come in a variety of sizes from table-top, hand-powered units to free-standing, hydraulically actuated models. They work by clamping the sheet metal in place, then applying force at the point of the desired bend and guiding the bend using a hinge. Most of these brakes are limited to 16 gauge steel (0.0598 in or 1.519 mm).

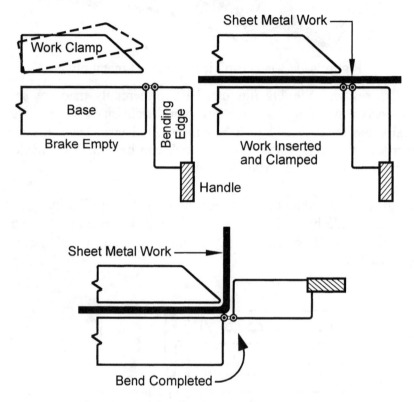

Figure 4–3. Sheet metal brake edge view.

- *Press brakes* are needed to bend heavier sheet metal gauges and all thicknesses of plate. Most are hydraulic, but some use an electric motor-driven flywheel. Much higher forces are needed to bend thicker metal and because of this, press brakes use a mating die design instead of the hinged designs of sheet metal brakes, Figure 4–4.

Medium capacity press brakes have 25- to 100-ton capacities. Shipyards use even larger press brakes with 2,000-ton ratings. By replacing the dies with long steel cylinders, smooth, long-radius bends can be put in thick steel plate for ship hulls, Figure 4–5.

Welding Fabrication & Repair 105

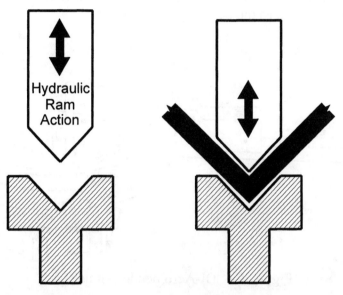

Figure 4–4. Press brake uses matching dies to make bends end view.

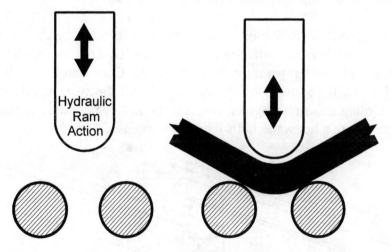

Figure 4–5. Roll brake bending of ship hull plate, end view.

- *Bending machines* come in dozens of sizes and designs. The smallest are hand- or air-powered, and the largest are hydraulic.

The Di-Acro Model 1A hand-operated bender forms round steel rod, square steel bar, flat steel bar, and steel tubing, Figure 4–6.

Heavy steel sections are heated in gas-fired ovens, with torches, or by induction heating to soften the steel. Heating lowers the metal's yield strength, and reduces the bending force needed.

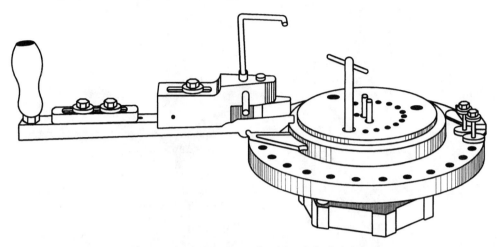

Figure 4–6. Di-Acro bender Model 1A.

When bending rolled shapes, pipe, and tubing, the forming dies must fit the work tightly. If the work is not supported and prevented from moving during the bending operations, results will be poor.

Cross-shaped or X-shaped extrusions that cannot be supported and confined by bending dies can be bent by casting the work *inside* a rectangle of low temperature metal alloy like CerroBend® which melts at 158°F (70°C), Figure 4–7. The rectangle with the work inside it is bent with tooling for a rectangular bar, and the low-temperature melting alloy is removed in a hot water bath. The alloy is reused.

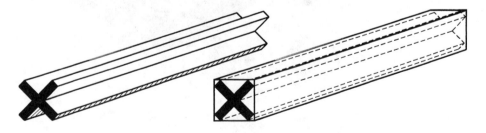

Figure 4–7. Extrusion to be bent (left), and extrusion cast inside low-temperature alloy to support it during bending (right).

Figures 4–8 through 4–15 show how to use Di-Acro benders to bend a variety of shapes.

WELDING FABRICATION & REPAIR   107

Circles can easily be formed with benders, but "spring back" must be taken into consideration. Most materials will spring back slightly after being bent. To compensate for this, it is necessary to use a radius collar with a smaller diameter than that of the circle required. Actual size can best be determined by experiment as the spring back varies in different materials. Materials should be precut to exact length before forming.

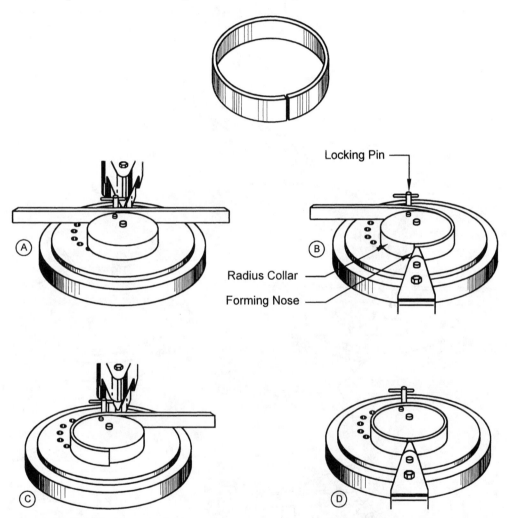

Figure 4–8. Forming a circle: (a) Set forming nose against material and clamp material against radius collar with locking pin, (b) Advance operating arm until forming nose reaches extreme end of material, (c) Relocate material and clamp with locking pin at a point where radius is already formed, (d) Advance operating arm until forming nose again reaches extreme end of material.

A sharp, zero radius bend can be formed using a zero radius block such as the one illustrated in Figure 4–9. The illustration shows the forming of strip stock, but this operation can be performed equally as well on round, square, and other solid, ductile materials. When forming heavy materials to a zero radius, their ductile limits must be taken into consideration. Providing a small radius on the bending edge of the block will avoid fracture or marking on the inside of the bend. By using a built-up forming nose and mounting two or more zero radius blocks on top of each other, the forming width capacity can be increased considerably.

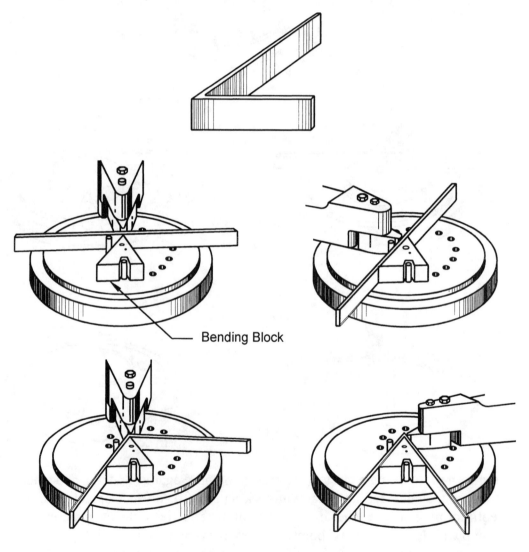

Figure 4–9. Making a zero radius bend.

Welding Fabrication & Repair

Scrolls and other irregular shapes can be formed in rigid materials even though the forming nose of the bender revolves in a perfect circle. This type of forming is done by using a collar having the same contour as the shape desired, Figure 4–10. This is accomplished by adjusting the forming nose so it is located only the material thickness away from the 'high point" of the contour collar. As the material will only bend where the contour collar offers resistance, the forming nose can lead the material around until it contacts the high point and exerts sufficient pressure to force it into the shape of the collar.

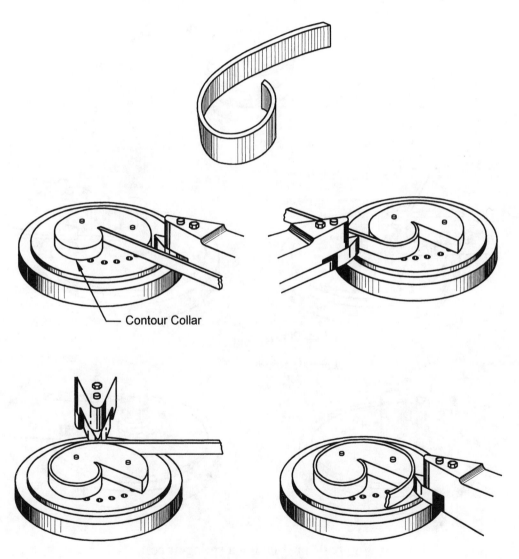

Figure 4–10. Scroll forming.

A spring or coil can be made in all materials by following the method illustrated in Figure 4–11. The number of turns in the coil is limited only by the height of the forming nose and the radius pin. Springs and coils can easily be made in any reasonable dimensions depending on the size and ductility of the material. The forming nose must be set so it will clear the end of the material held by the locking pin. The maximum length of this end is determined by the ductile limits of the material as it must be sufficiently rigid so it will not bow between the forming nose and the radius pin.

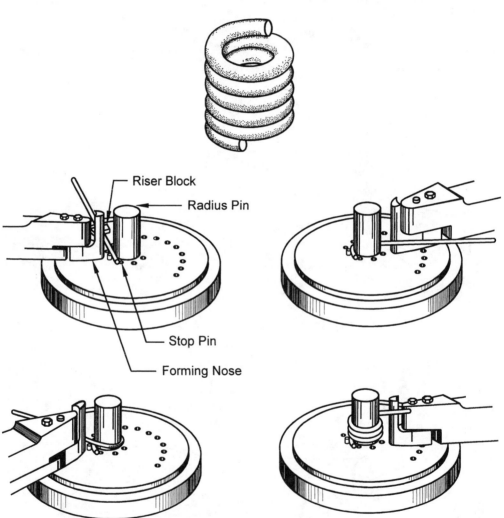

Figure 4–11. Making a spring or coil.

Welding Fabrication & Repair

The forming roller method of tube bending is recommended for all large bends where the center line radius is at least four times the outside diameter of the tube, Figure 4–12. It can also be successfully employed for bending pipe or heavy wall tubing to smaller radii and is the most practical method of bending small diameter tubing. The forming roller and radius collar must be grooved to exactly fit the tube. The tube must not be allowed to slip during the bending operation as even a slight amount of slippage will cause distortion. Nearly all pipe and tube bending methods trap and confine the workpiece in this manner.

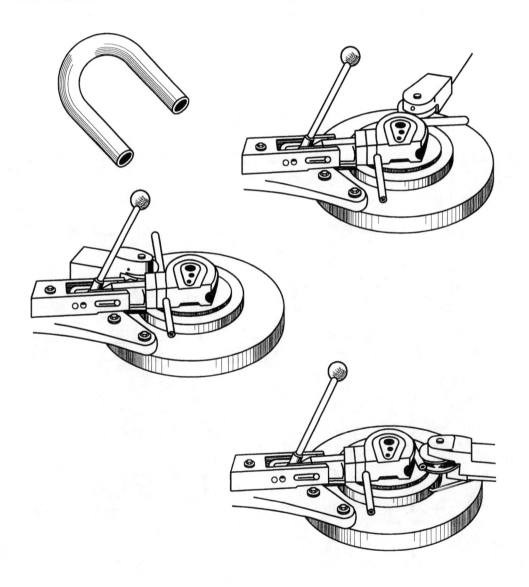

Figure 4–12. Forming roller method of tube bending.

Angle iron with flanges or legs out can be formed with benders, but this type of bending presents problems because stresses set up within the material often cause the workpiece to twist out of plane after forming, Figure 4–13. The twisting is generally more pronounced in fabricated angle than in standard mill rolled angle. When the flange bent edgewise is less than half the width of the vertical flange, the tendency to twist is greatly reduced. As the dimensions of the angle and radius of bend vary with almost every job, it is impractical to offer a standard group of accessories for this type of forming. The user must prepare these parts. The flange should be closely confined in the radius collar; as little as 0.002 to 0.003 inches (0.051 or 0.0.076 mm) variation in clearance can make a great difference in bend quality. Minimizing and controlling clearance is especially important in thinner materials.

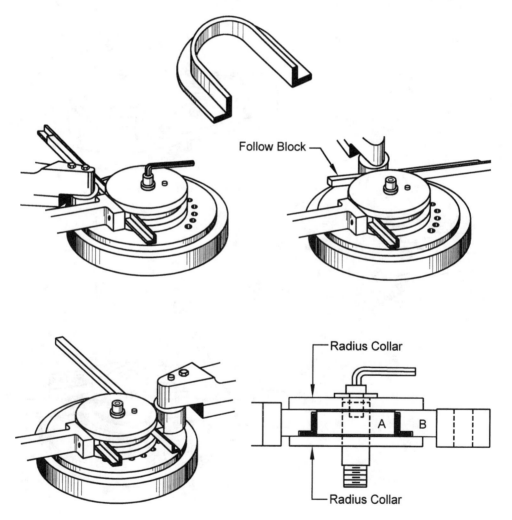

Figure 4–13. Making a U-bend in angle iron flanges or legs out.

WELDING FABRICATION & REPAIR

Benders will form channels provided the flanges are adequately confined during the bending operation so they will not buckle or distort, Figure 4–14. Usually the radius of the bend should be three to four times the width of the flange to allow for stretching of the metal. Although this ratio is primarily determined by the thickness and ductility of the material, it can often be reduced considerably. As the different dimensions of the channel shape vary with almost every job, the user must prepare these parts for his job. Minimizing flange clearance is important just as it is in angle and channel bending.

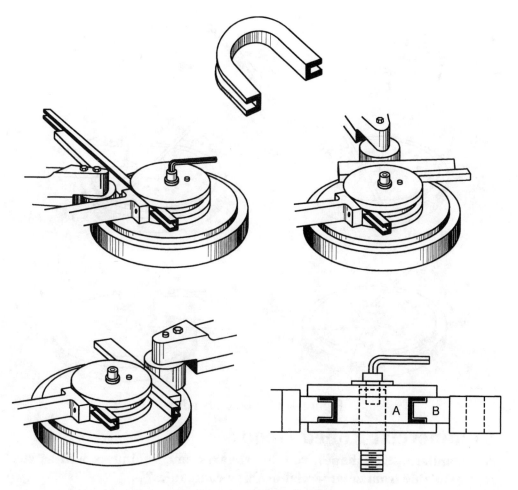

Figure 4–14. Channel bending legs out.

114 CHAPTER 4   BENDING & STRAIGHTENING

Forming zero radius bends around square, rectangular, or other multi-sided blocks employs the same principle used in scroll bending. The forming nose "leads" material between corners of the block, Figure 4–15. Any number of zero radius bends can be obtained in one operation by this method in all types of solid materials. This method of bending is limited by the size of the square block and the ductility of the material. In general, when squares larger than 1 inch are needed, they should be formed in progressive operations using the zero radius block shown in Figure 4–9 for zero radius bending.

Figure 4–15. Square bending.

Commercial Rolled Goods

Are standard steel shapes, rounds, squares, and tubing as well as steel pipe available from commercial bending companies?

Yes, a wide variety of steel shapes formed into rings (or sections of rings) are available on short notice from commercial sources. Figure 4–16 shows the standard mill shapes available from Chicago Metal Rolled Products Company. Both the maximum sizes and common descriptions are shown. Other metals can be ordered as well.

Welding Fabrication & Repair 115

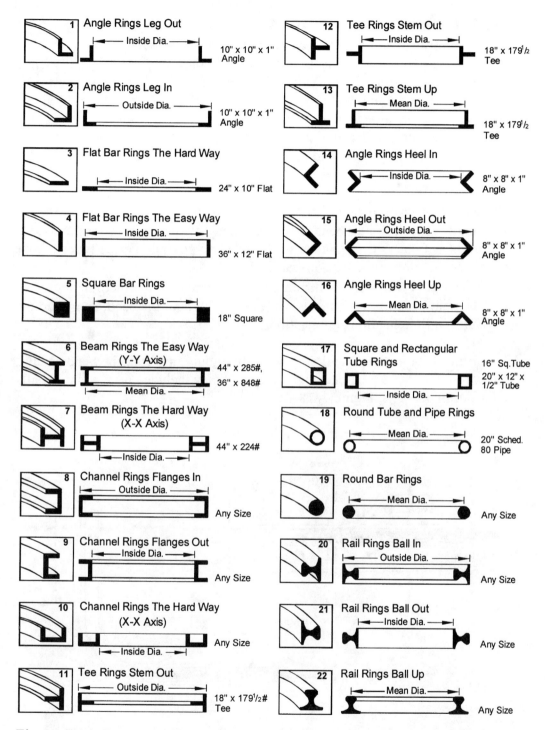

Figure 4–16. Commercially available steel rolled shapes and maximum sizes.

Other Bending Methods

What is mandrel bending and what are its advantages?

Mandrel bending is the term used when a mandrel, or former actually goes *inside* the pipe or tube being bent. It is held there by a mandrel support rod, and supports the tubing wall to keep it from collapsing, Figure 4–17. Mandrel bending requires a machine with a longer bed than conventional rotary bending machines, and a frame strong enough to resist the mandrel forces. Lubrication of the inside of the pipe or tube is needed to reduce forces on the mandrel and often requires removal later.

Three advantages of mandrel bending are that it can:
- Create a much tighter bend than those made without an internal mandrel; bends as tight as the diameter of the tube are possible versus two to three diameters of the tube for non-mandrel methods.
- Make tight bends without bending wrinkles common in other methods.
- Maintain the workpiece inside diameter throughout the bend, an important factor in fluid handling applications.

The form of the mandrel will vary with the tubing material, the bend radius, and wall thickness. Mandrel designs vary and may include plugs, a single ball, a disc, or multi-balls, Figure 4–18.

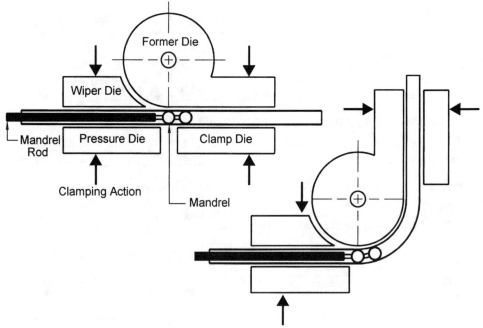

Figure 4–17. Mandrel bending.

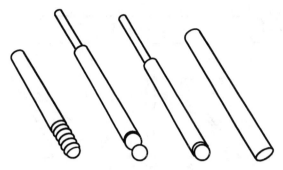

Figure 4–18. Typical mandrel designs.

How can sand prevent pipe (or tubing) walls from collapsing during bending?

When bending soft materials, with or without a bending machine, the pipe walls tend to collapse. Filling the pipe or tube interior with dry sand and capping the ends will avoid or reduce this problem, Figure 4–19. Be sure to leave small holes in the plugs to allow any steam generated by the sand to escape if heat is applied. The sand needs to be well tamped before the ends of the work are capped.

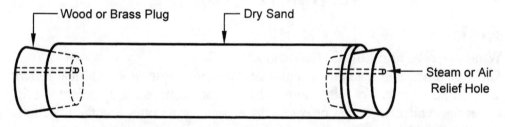

Figure 4–19. Using sand to prevent pipe wall collapse while bending.

What is wrinkle bending?

It is an old method of bending medium to large steel and copper pipe without a bending machine. This method is useful when an elbow fitting of the proper size and angle is unavailable.

Wrinkle bending is performed by heating a band of the pipe about halfway around its circumference on the inside of the desired bend. A rosebud torch tip is essential, and on pipe 8-inches (20 cm) diameter and larger, two rosebud torches are needed. When the band is dull red, apply a bending force, usually with a come-along. This will upset the metal on the inside of the bend. Each wrinkle should bend the pipe between 9° and 12°. Nine or more wrinkle bends are needed to make a right angle bend. Because the wrinkle-metal also upsets the outside of the pipe, there is very little reduction to the pipe diameter. See Figure 4–20.

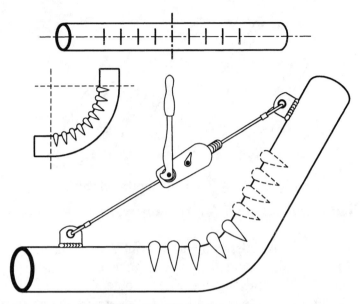

Figure 4–20. Wrinkle bending.

Section II – Rolling

Rolls

What are rolls and how do they work?
Rolls are a bending machine design that uses three or more cylindrical rollers to form sheet metal or plate into cylinders or sections of cylinders. Rolls, sometimes called *slip rolls* or *roll benders*, can form cones and their sections too. Small table-top, hand-powered versions are widely used for sheet metal in HVAC. Factory-installed machines as big as a house, driven by 100 horsepower motors, bend heavy steel plate for water and petroleum storage tanks, boilers, and ship hulls. Roll benders are made in every conceivable size.

Figure 4–21 shows how rolls work. One end of the roll bender drops out of the way to allow the work to be withdrawn after rolling is completed. Figure 4–22 shows a table-top, hand-powered roll bender.

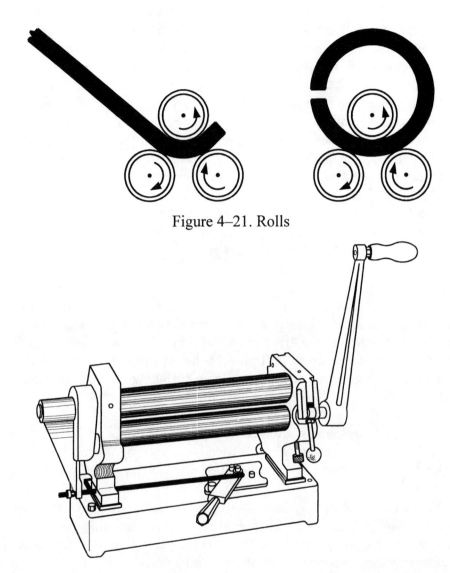

Figure 4–21. Rolls

Figure 4–22. Table-top slip rolls from Di-Acro Incorporated.

Section III – Heat Forming

Flame Bending, Shrinking & Straightening

What are the differences among *flame bending*, *flame shrinking*, and *flame straightening*?

These terms describe three slightly different activities that all use the effects of heat, thermal expansion, partial constraint, and upsetting to permanently change the shape of metal, usually carbon steel.

Flame bending is often used in shipbuilding. It has the ability to put curves into steel plate without a hydraulic press. Only an oxyfuel torch and a water source are necessary. One or more curves can be made in a plate; they can be gradual, or severe. The curves can be simple, on just one axis, or complex. Curves with changing radii of curvature are easily formed. Flame bending can even form a plate into a concave shape. If too much of a bend is applied, flame bending can reduce or remove the bend. In shipyards, flame bending on plate or panels is sometimes called *line heating,* because heat is usually applied along straight or curved lines. Steel fabricators also use flame bending to put camber (vertical curve), or sweep (horizontal curve) into I-beams. Other steel rolled sections, pipe, and tubing can be flame bent too. In fact, any metal can be flame bent. The larger the coefficient of thermal expansion the metal has, the better flame bending works.

Flame shrinking uses the same physical processes as flame bending, but accomplishes a different result. It *removes* wrinkles in steel plate panels after welding operations. Wrinkles also form around holes cut in steel panels, and flame shrinking can remove these as well. This is a common shipyard activity. Without this shrinking, every hull plate in the ship would carry the outline of the supporting steel frame behind it. Flame shrinking on plate is sometimes called *panel shrinking.*

Flame straightening employs the same techniques as flame bending, but is used to straighten steel members instead of bending them. While flame straightening can be used to remove bends in pipe and small rolled sections, its principal application is the in-place repair of damaged, heavy steel sections: beams, columns, girders, trusses, and crane rails. Any steel member can be repaired by flame straightening. These repairs are often done on bridge beams struck by over-height loads, or damaged by over-weight vehicles. Sometimes bridge columns need straightening after being hit by run-away vehicles, loads, or ships. Steel erection mishaps, structural steel fire damage, and cargo container crane boom collapses are also excellent candidates for flame straightening. Flame straightening is also very useful for straightening railings, racks and other long structures that have been bent by welding shrinkage.

Flame straightening and bending can be performed using only thermal expansion forces to bring the metal into position. But often hydraulic jacks, wedges, or come-alongs are used to apply additional external force, and to increase the deflection during each heating and cooling cycle.

What are the advantages of flame straightening of damaged structures?
Flame straightening has three significant advantages:
- Cost savings of 25% over direct repairs are typical.
- Repairs are usually completed in a few days, or a few weeks and there is no waiting for the fabrication of new structural elements.
- Equipment required to make the repairs is relatively simple and available in remote areas and third world countries; repairs of heavy sections are often made without the use of additional heavy equipment.
- Strength of the steel is unaffected by flame straightening.

A handful of expert flame straightening contractors in the US perform their skills world-wide, and often work twenty-four hour shifts to complete critical repairs.

What is the principle behind all these steel-shaping operations?
Four physical events must occur in the following order. There must be:
1. Local and rapid heating leading to thermal expansion caused by the application of an oxyfuel torch. More than one torch may be used.
2. Constraint from the cooler and surrounding unheated steel on the hotter metal; this is because the cooler metal has expanded less than the metal in the heated area.
3. Localized upsetting, that is when the heated metal expands and if the thermal expansion exceeds the yield point of the metal, plastic deformation occurs in the *unrestrained* direction, so the metal is thickened.
4. When the metal cools, it shrinks in all three directions and causes a geometric change from the original shape.

See Figure 6-43.

Changing the shape of steel relies upon rapid thermal expansion, because if the entire workpiece were gradually brought up to temperature, it would expand and contract evenly, retaining its shape. To keep the heating and expansion uneven, water or sometimes a compressed air and water mist is applied as soon as the upsetting occurs. Water is applied with a hose, or in a mist with the atomizer shown in Figure 4-23. These sprayers may be shop-built. The advantage of using a mist is that less water ends up in puddles on the floor than when water is applied with a hose.

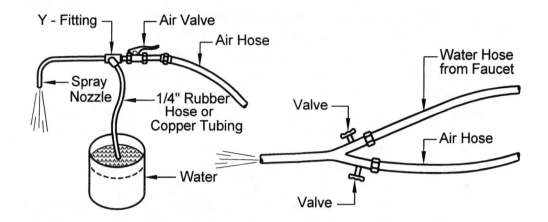

Figure 4–23. Compressed air and water mist sprayers for flame shrinking.

The heating and cooling cycle is repeated as soon as the metal has cooled to room temperature. In straightening operations, hundreds of cycles may be needed.

Heat may be applied in spots, rings, lines, or triangles to get the required shape changes in the metal. The proper application of heat and cooling can restore previous shapes or create new ones.

Today these three processes depend on a skilled craftsman's experience and judgment, but development work is under way to automate flame bending for shipyard operations. Soon a computer will calculate the heated areas needed to develop the shapes required in the steel plates and will guide oxyfuel torches to do the job.

Why don't flame straightening, flame bending, or panel shrinking heat weaken the structural steel members?
The American Welding Society codes for bridges and structures limit flame straightening temperatures to 1200°F (648°C) because it is generally recognized that below this temperature little or no permanent weakening of the steel occurs. At this maximum allowed temperature, steel is a dull, cherry red in subdued light. To insure that the work is not overheated, temperature indicating crayons or electronic instruments should be used when working in bright sunlight.

Flame Bending Bars
How is a rectangular steel bar flame bent?
The steps are:

WELDING FABRICATION & REPAIR

1. Apply heat to the side of the object you wish to shrink. An easy way to remember this is that "the hot side is the short side."
2. For a rectangular metal bar, maximum bending is usually obtained with a 30° triangle that runs from the outside edge of the "short" side seven-eights of the way down the bar's width, Figure 4–24 (a). The cool, unheated metal in the bottom one-eighth of the work on the inside of the bend acts as a hinge. Use weldor's chalk to mark the triangle. Initially use a 10° or 15° pie-shaped segment to get started, and work up to wider angles when you see the results.
3. Use a neutral oxyfuel flame and stay within the marked triangle. Move the torch continuously along the heating path slowly enough to heat the metal under it to a dull red (in subdued light, not direct sunlight) before moving on. Do not exceed 1200°F (648°C).

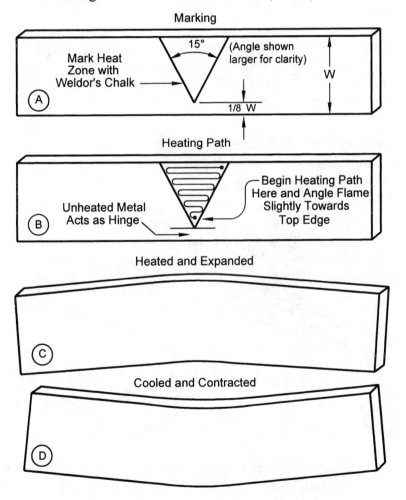

Figure 4–24. Flame bending a rectangular steel bar.

4. Choose an oxyfuel torch tip that will heat the work quickly, yet stay within the chalked lines. A rosebud tip and two or more torches are needed for large work. Both sides of thick material must be heated.
5. Because it is impossible to visually regulate metal temperature in bright sunlight, use a temperature-indicating crayon, or non-contact hand-held electronic pyrometer to keep below the 1200°F temperature.
6. When using water, or a compressed air and water spray to cool the work, begin by cooling the area *around* the heated zone. On smaller jobs such as this example steel bar, even wet rags can be used for cooling. We want to maintain the maximum temperature differential *between* the hot and cold zones.
7. When working with higher carbon content alloys, do not apply the cooling spray to the heated zone directly. You run the risk of inducing hardening.
8. Begin with a small trial area, say 10° to 15°, to gauge the shrinkage and avoid overshooting the mark. Use wider V-shapes, up to 30°, when greater bending is needed. It is much better to perform several small heats than to over-bend and have to use additional heating/cooling cycles to correct your mistake.

Flame Bending Rolled Shapes

A steel or aluminum rolled shape must be bent, but no hydraulic bending equipment is available. How is this done?

Apply torch heat along the paths shown in Figure 4–25, then cool with water or a water and compressed air mist. Do not apply the water or mist directly to the heated metal. Begin by cooling the dark, unheated area around the dull red metal and work inward to the hotter metal. More than one heating and cooling cycle will probably be needed.

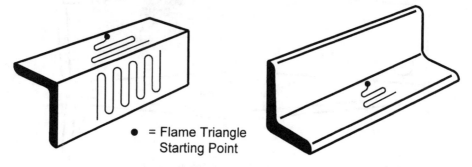

● = Flame Triangle Starting Point

Figure 4–25. Using torch heat to bend shaped members. Lines show torch path and heated areas. (continued on next page)

WELDING FABRICATION & REPAIR

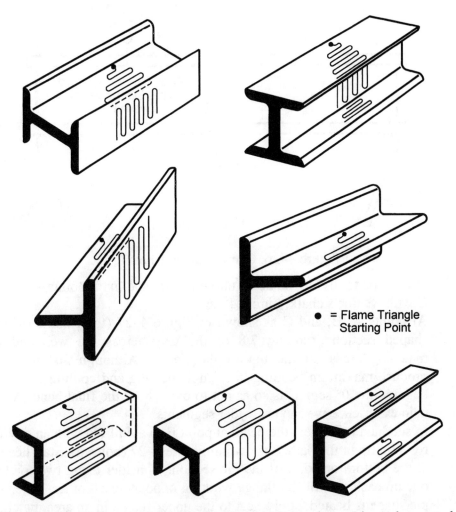

Figure 4–25. Using torch heat to bend shaped members. Lines show torch path and heated areas. (continued)

Cambering I-Beams
What is the proper procedure for cambering an I-beam?
Follow these steps:
1. Sight along the edges of the flanges to see if the beam already has a camber. If camber exists, then plan on increasing the existing camber.
2. Place the beam on horses or other supports so that it is supported only on the ends. Place a third horse or support under the beam exactly at the beam's midpoint. The beam and its center support must not touch. Measure the distance between the bottom of the beam and the top of the center support and record this distance on the web of the beam above this point for later reference, Figure 4–26.

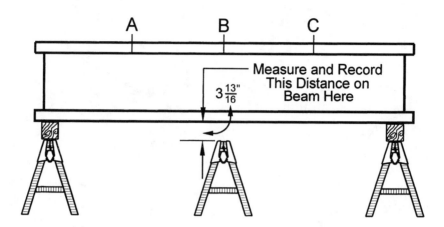

Figure 4–26. Cambering an I-beam.

3. Use a tape to divide the beam into quarters, and mark points A, B, and C with weldor's chalk, Figure 4–26.
4. At points A, B, and C as shown in Figure 4–27 (top), mark 10° pie-shaped sections running 7/8 of the way across the web and the matching areas on the top of the flange. Although 30° segments produce maximum bending in a single heating and cooling cycle, we begin with 10° segments so as not to over shoot the final bend. As we gain experience we can use wider segments.
5. In subdued light, heat the beam at point B to a dull red with an oxyfuel rosebud tip in the pattern shown in Figure 4–27 (bottom). Start heating at the bottom or point of the pie section. Consider using two torches, one on each side of the flange working opposite each other. When the pie shape is heated, apply heat to the upper flange in an area matching the top width of the pie shaped-section. Cool the beam with water starting from the cooler, darker black areas around the heated areas and work into the red heat zones. When the beam is back at room temperature, measure the camber change at B. If more camber is needed, and it probably will be, mark the pie segment and top web area to be heated and, apply heat at points A and C in a similar manner.
6. Check camber at A, after heating and cooling at B and C. If more camber is required, apply heating and cooling at D and E. D is midway between A and B; E is midway between B and C.
7. If more camber is needed, repeat at additional intervals midway between previously heated zones. Avoid heating the same areas twice as this will not lead to a smooth, sweeping bend.

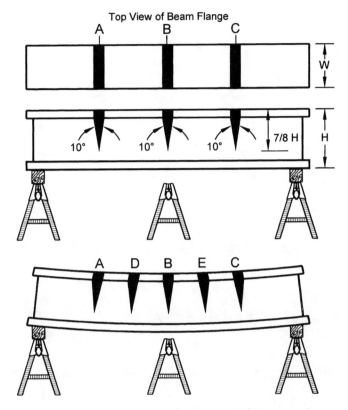

Figure 4–27. Cambering an I-beam.

Line Heating

How does *line heating* differ from other operations where heat is used to bend metal?

Line heating is flame bending performed on steel plate, rather than rolled shapes, pipe, tube, or sheet metal. While line heating can be used in a variety of steel fabrication activities, shipbuilding is its principal use. Steel hull plates vary in size but are typically 6 by 21 feet (2.3 by 6.5 meters). Hull plate can run from about 1/2 inch to several inches in thickness.

In the simplest instance, if oxyfuel torch heat is applied along a line in a plate, the plate will contract into a slight V-shape when it has cooled, Figure 4–28 (top). If a series of parallel and evenly spaced heat lines is applied, the plate curves evenly across its width, Figure 4–28 (center). Adding more heat lines closer together, further increases the bend in the plate, Figure 4–28 (bottom).

Unbalanced shrinkage forces between the heated and unheated sides cause angular distortion just as in other heat bending applications.

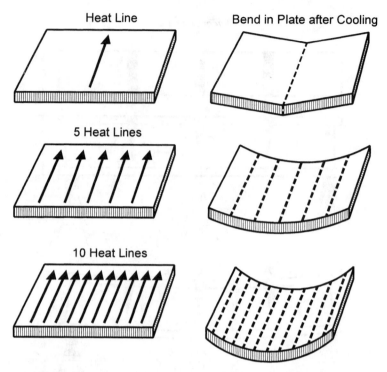

Figure 4–28. Effect of a single heat line on a steel plate (top), effect of multiple heat lines (center), and adding even more heat lines (bottom).

What is the typical spacing between heat lines?
See Table 4–2.

	Heat Line Spacing	
	Inches	mm
More plate curve	4	100
↓	6	150
↓	8	200
Less plate curve	10	250

Table 4–2. Typical heat line spacing.

In general, how much deflection on a plate will a single line heat produce under shipyard conditions with the proper equipment?
The range of deflection from one heat line runs between 3/4 and one degree for a single heating and cooling cycle.

How far should the torch flame be from the steel surface?
With the larger, multi-orifice oxyfuel torches used for line heating, a tip-to-plate spacing of between 3/4 and 1 inch (20 to 25 mm) will put the hottest part of the flame on the steel and create the maximum bending action.

How do travel speed and path shape influence how much deflection occurs?

There is an optimum line travel speed for maximum deflection; too slow or too fast torch travel will reduce deflection. Weaving serpentine lines, instead of straight-line travel, increases deflection in a 1-inch plate almost 20%.

How fast should the torch move along the heating line?

It should move rapidly enough that the metal temperature does not exceed 1200°F. The exact speed depends on the torch, tip, fuel and oxygen pressures, but 1/2-inch (12 mm) steel plate should see a minimum torch speed of 15 inches/minute (400 mm). Thicker steel in the 1- to 2-inch (25-50 mm) range should not fall below 10 inches/minute (250 mm/min). These figures are approximate.

When only a small curvature adjustment is needed, line speed may be double the normal speed.

How is plate temperature measured and regulated?

Temperature regulating crayons work well to measure maximum metal temperature, but the experienced craftsmen tend to watch the color of the steel in subdued light to control temperature. Non-contact, optical pyrometers can also be used.

How are temperature-indicating crayons used to measure maximum metal temperature?

The oxyfuel heating torch is aimed in the direction of torch movement. The temperature-indicating crayon is marked on the steel on the line-heating path about 2 inches (50 mm) behind the torch. If the plate marking turns color from its initial one as shown in the manufacturer's tables within 2 seconds, the calibrated temperature exists on the plate. Tests have shown that at this 2-inch following distance, the steel's temperature falls about 270°F (150°C) from the under-torch temperature. To get the metal temperature under the torch, we must add back the 270° to the temperature of the crayon.

How deep should torch heat penetrate to cause the maximum bending?

Maximum bending takes place when the line heat penetration reaches the middle of the plate. However, this also causes maximum transverse shrinkage (across the plate, perpendicular to the heat lines), so the best compromise between shrinkage and bending is to have line heat penetrate to just short of the center of the plate.

What would happen if two torches were to apply the same heat to both sides of the plate?
Balanced contraction occurs. There is no bending, but there is shrinkage after cooling. By heating triangular areas on both plate sides at once, shrinkage forms the plate into a concave shape. More on this follows.

How does the maximum temperature on the heat line affect bending?
The higher the temperature, the more bending occurs. However, we want to limit the maximum temperature to avoid changing the properties of the steel and this turns out to be about 1200°F (650°C) for most ship plate steel.

A cooling water stream or a mist of water and compressed air is usually applied following the torch along the heat line. Why is this done?
There are several reasons:
- Cooling maximizes bending by confining the heat to the heat line, and maintaining the maximum temperature differential between the heated and the unheated areas.
- The water mist keeps the entire plate at a lower temperature and allows the heat bending cycle to be repeated quickly. However, if only a small bend is needed, cooling water or mist can be omitted.

How much cooling water is typically applied by hose and how far behind the torch should the water or cooling mist be applied?
Cooling water is applied about 5 inches (120 mm) behind the torch and at the rate of about 3/4 gal/min (3 l/min).

How closely can a skilled line heating craftsman match the templates?
Roughly ±0.2 inches (±5 mm).

If too much curve or bend is applied to a plate, how can this process be reduced or reversed?
Flip the plate over and apply line heat in paths parallel to the initial heat lines.

What other steps can increase bending during line heating?
Applying compression stress to the heated side of the plate can double angular deflection. This is a common practice in line heating. Wood and steel wedges pull the plate up on one side and cleats hold it down against the work surface on the other, Figure 4–29. Sometimes a crane applies stress to the work.

WELDING FABRICATION & REPAIR 131

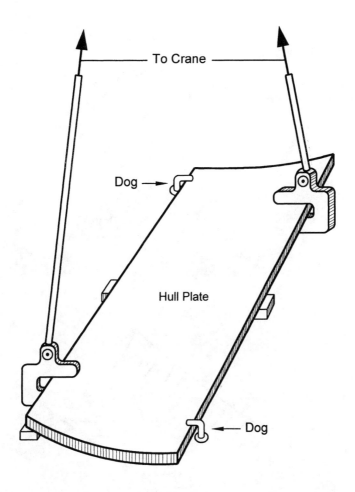

Figure 4–29. Using restraint to increase bending.

What line heating pattern makes a plate concave (dished)?
Use heat triangles, Figure 4–30. Heat triangles along the edges of the plate will shrink the edges, leaving the center of the plate un-shrunk, and so makes the plate convex.

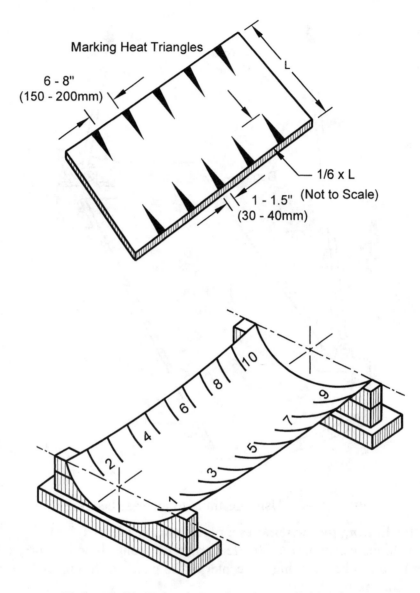

Figure 4–30. Using heat triangles to dish a plate.

What equipment is needed to perform line heating on large steel plates?
- Crane for handling the plates.
- 4" × 4" wood blocks to prop up corners of the plate.
- Assorted wood wedges.
- Steel wedges 6 inches (15 cm) long and 1.5 inches high (40 mm).
- Dogs to secure plates to the work platform.
- Large pry bar to move the workpiece and wedges.
- Sledge to tighten wedges under the work.

WELDING FABRICATION & REPAIR

- Multi-orifice oxyacetylene or oxypropane torch with fuel and oxygen regulators, flow meters, and compressed gas sources.
- Temperature indicating crayons 700 to 1300°F (400 to 700°C).
- Assorted marking and measuring tools.
- Metal or wood templates to gauge plate curvature.
- Cooling water hose or compressed air-cooling water atomizer assembly.
- Safety equipment—safety glasses, tinted face shields, and gloves.

Why not just use bending rolls and avoid line heating?
Bending rolls form plate into a cylinder (or sections of a cylinder) with a *single*, constant radius. Line heating can produce bends in plates with *changing* radii. A gentle bend can transition to a severe bend and there may be several changes in curvature.

Line heating to a great degree can replace bending rolls, but not the other way around. When rolling capacity is limited, line heating can be used.

It is also possible to begin some bending jobs on rolls and then use line heating to add the final adjustments.

What templates are used to check that the work has the desired curves?
There are several different designs:
- Pattern templates cut from plywood.
- Reusable templates that can be formed to a given curve and lock the curve in place, Figure 4–31 and 4–32.

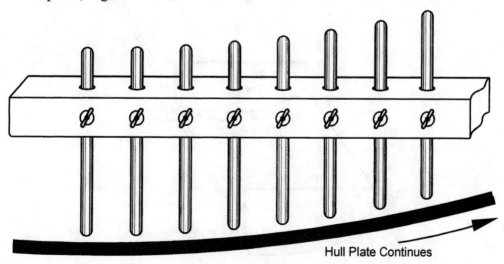

Figure 4–31. Templates used for checking line plate curves.

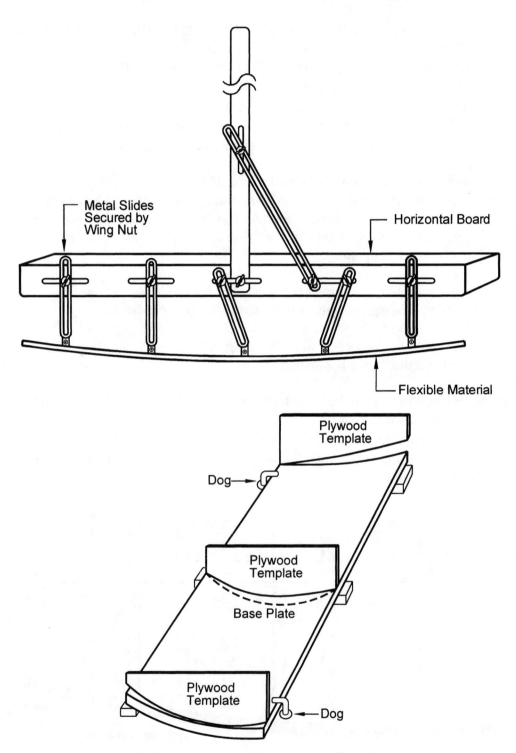

Figure 4–32. Templates used for checking line plate curves.

Author's Note: Line heating is based on well understood scientific and engineering principles, but the craftsmen who routinely use line heating to form ship hull plates are artists. Based on their experience, they examine a pattern for a complex plate and know where the heat lines should go and how fast to move the torch. Although it takes several years of experience to excel at this craft, using the basic concepts given here allows the beginner to make simple curves and shapes in steel plate.

Flame Straightening Large Pipe

You are installing a length of 20-inch (50 cm) pipe with a flanged end. The pipe has a 5/16-inch (8 mm) thick wall. The pipe has a slight bend in it preventing the flange from making up properly. How can this job be completed using this pipe?

Use flame straightening to remove the kink in the pipe. Alternately, if the pipe was straight and a kink was needed, it could be added with a flame bending operation.

To remove this kink, heat a 3½-inch diameter area to a dull red and cool, Figure 4–33. Repeat if necessary. You can undo any straightening step by heating on the opposite side.

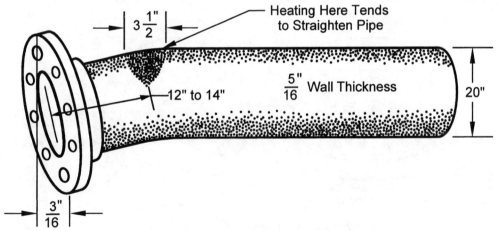

Figure 4–33. Flame straightening pipe.

Flame Shrinking Procedures

You used an oxyfuel flame to cut a circle out of steel plate. When it cooled to room temperature, the circle was no longer flat: it bulged in the middle and was now dish-shaped. What happened and how can we make the circular plate flat again?

Cutting torch heat expanded the donut shaped area around the edges of the circular plate. Because this outer ring of hot steel was constrained from expanding freely by the cooler metal in the center of the plate, thermal expansion forces exceeded the yield strength of the steel. This outer donut of steel permanently stretches to fit the existing cooler inner area. Upon cooling, the outer ring then shrinks more than the inner (remember the inner ring is already cool), squeezing the inner excess material into a dish shape. See Figure 4–34 (a).

To get rid of the excess material in the center of the circle we merely reverse the process: apply torch heat to the center of the circle, let the cool outer material contain the hot inner material, and cause it to thicken. When this center material cools, it will shrink and remove the dome of excess material, Figure 4–34 (b). Often several heating and cooling cycles are needed. To speed this process and to prevent the entire disk from getting hot when we want only the central portion to heat, the water spray apparatus in Figure 4–23 is often used.

If the disk is not too large, the outer donut of steel can be stretched to let the larger inner circle flatten. This is done by hammering around the edge of the steel plate, but does not produce as visually attractive results.

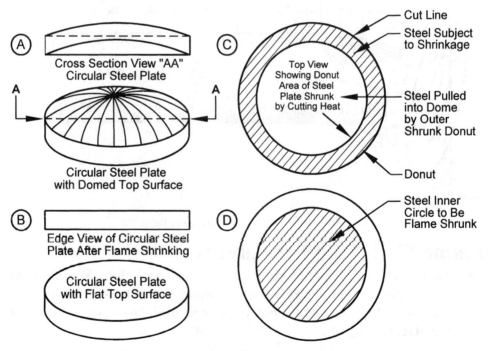

Figure 4–34. Flattening a circular disk.

You have welded steel panels onto a steel frame. Ripples now appear in the center of each of the steel panels, Figure 4–35. How do you cure this problem?

This is just like the excess material in the center of the steel plate. The only difference is the source of the heat causing the distortion. The cure is the same: apply oxyfuel heat to the center of the panel, and shrink it with a water spray. Several heating and cooling cycles are usually required. This is a common shipyard process; one weldor heats the panels with a rosebud tip and another follows with a water spray.

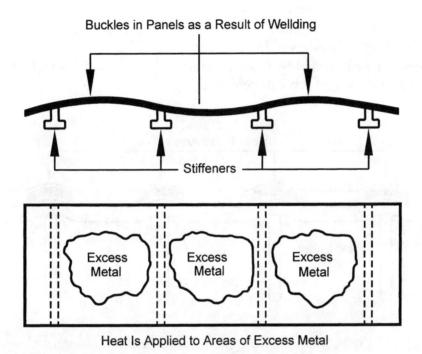

Figure 4–35. Ripples in center of welded panel.

Fence & Railing Problems & Solutions

A fence has been fabricated of pipe, tubing, or shaped steel. Welding shrinkage has caused the top rail to deform and pull down on each side of the vertical posts, Figure 4–36. How can waves in the top rail be straightened?

Flame straightening is the perfect solution. Apply an oxyfuel flame on top of the rail above the welds. Begin by heating an area about 2 inches (5 cm) long and extending about one-third of the circumference of the pipe or tube above each weld, Figure 4–36. Several heating cycles will probably be required. It is always best to perform several heats, than to overshoot the mark and have to fix the excess correction, so always begin in smaller areas until you see how much correction heating a given area produces.

Here are some other suggestions for avoiding problems building fences and railings.
- Fit the steel to minimize gaps that will be filled with weld metal, because the more weld metal, the greater the shrinkage.
- Minimize the size of the weld beads themselves.
- Beginning welding in the middle and working out toward the ends of the fence will produce best results.
- Heating the top of each fence rail over the welds either right before welding, or right afterward will reduce cambering.
- Bends in what should be straight members in excess of 1/8 inch (3.2 mm) in a 2-foot (60 cm) run will be noticeable.

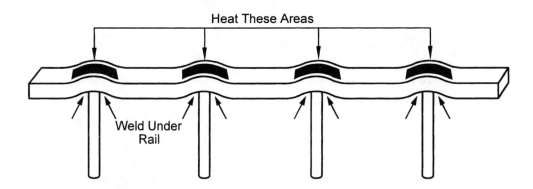

Figure 4–36. Fence weld distortion and its solution.

A cover plate must be applied to the bottom flange of an I-beam, Figure 4–37 (a). After welding on this plate, the beam and plate are no longer straight but form an arc or curve. See Figure 4–37 (b). How can we straighten this beam?

The heat of the welding has distorted the bottom flange, so we will apply oxyfuel heat to the top of the beam—the flange without the plate. By repeatedly oxyfuel heating and water spray cooling, the beam can be completely straightened.

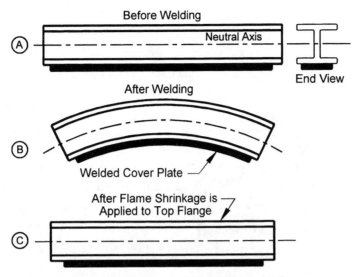

Figure 4–37. Flame straightening an I-beam.

Using Gravity to Control I-beam Camber

How can gravity be used to control beam camber when flange plates are applied?

When a plate is applied to the flange of an I-beam, we can increase or decrease the camber slightly by the placement of support for the beam, Figure 4–38.

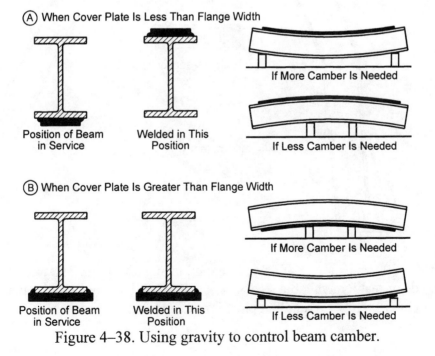

Figure 4–38. Using gravity to control beam camber.

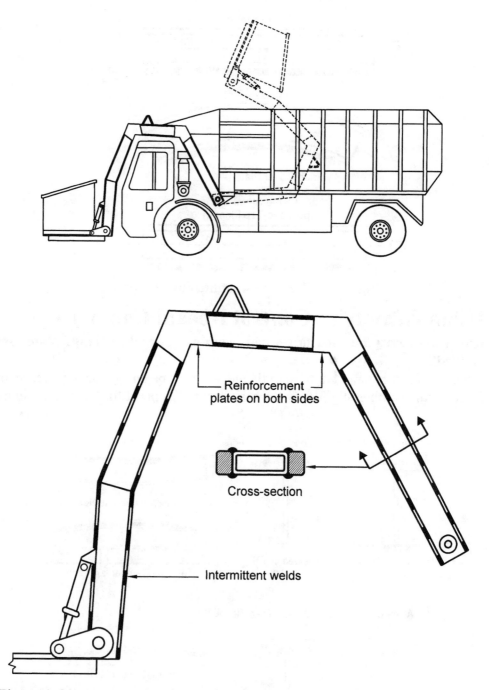

Figure 4–39. An example of a weldment replacing a forging. The truck boom is made of readily available steel stock: solid steel bars, rectangular tubing, and flat plate for the gussets.

Chapter 5

Vehicle Welding

It may be those who do most, dream most.
—Steven Leacock

Introduction
Repairing vehicles is one of the biggest welding applications. This chapter looks at four problems that vehicle welding presents: repairing cracked truck frames, attaching roll bars to unibody vehicles, welding up small holes in body sheet metal, and transferring a fishmouth shape to the workpiece.

Cracked Vehicle Frame Repairs
There are prominent labels on the C-channel frames of modern tractors and heavy trucks warning against cutting or welding on them. Why is this warning there?
To save weight, the manufacturers used thinner, lighter, steel C-channel members with a special heat treatment that provides extra strength. Welding and torch cutting on these members destroys the strength of the factory heat treatment. Do not weld, flame cut, or drill on these members if they have not failed. If you have to mount something on the frame, use the extra, unused existing holes put in at the factory. However, if C-channels must be repaired, minimize welding on them.

How should a cracked C-channel truck frame be repaired?
Use the following steps:
- Clean the repair area. First, steam clean and scrub the entire area surrounding the weld. (This cleaning is particularly important for waste hauling vehicles.) Then use an oxyfuel torch to dry this area and remove remaining mill scale. Finally, wire brush the area down to shiny metal.
- Compare your failure with those shown in Figure 5–1 to determine which case your frame failure best matches and then follow the repair steps for that case.

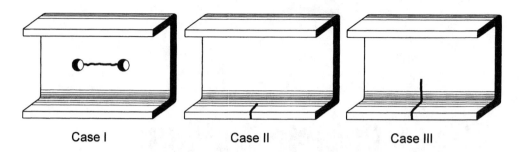

Figure 5–1. Typical truck frame cracks.

Case I—Horizontal crack along the web between factory-drilled holes.
This is a common case and cracks as long as 10 inches occur. See repair steps in Figure 5–2.

- Grind a V-groove to within 1/16 to 1/8 inch of the thickness of the web steel along the path of the crack and extending 2 inches beyond the initial crack on each end. Put this V-groove on the inside of the C-channel, Figure 5–2 (b). Use a copper backing plate clamped to the back of the groove to protect the back of the weld from atmospheric contamination.
- Using SMAW low-hydrogen electrodes, fill the V-groove with weld metal and grind it flush. Use enough current for a full penetration weld. See Figure 5–2 (c).
- Grind the weld flush (on the outside of the channel too if necessary to make it flat) making sure to leave the ground surface as smooth as possible. Any irregularities or scratches are stress raisers.
- Cut and fit a ½-inch thick carbon steel reinforcement plate on the inside of the web extending at least 6 inches beyond the ends of the weld repair. Grind this plate on the lower and upper edges so it fits *tightly* against the web and edges of the flanges for its entire distance, Figure 5–2 (d).
- Using existing factory-drilled C-channel holes if possible, secure the reinforcement plate to the web with bolts matching the diameter of these holes. Holes are usually 1/2-, 9/16-, or 5/8-inch diameter. Use a washer under each nut. If no holes are available, drill your own.
- Stop. The repair is complete. No additional welding is needed. See Figure 5–2 (e).

Welding Fabrication & Repair 143

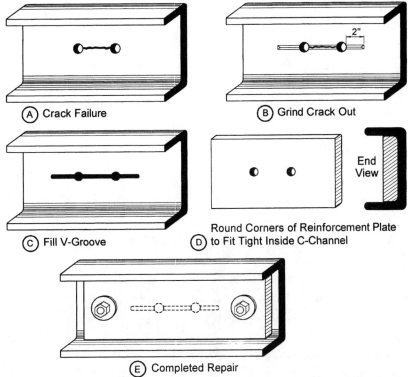

Figure 5–2. Case I: Horizontal crack in truck C-channel between factory-drilled holes in web. This is a common case and cracks as long as 10 inches (250 mm) can occur.

Case II—Crack on bottom flange perpendicular to web only.
- Here are the repair steps:
- Grind a V-groove half way through the thickness of the flange along the path of the crack. See Figure 5–3 (b).
- Using SMAW with a low-hydrogen electrode, fill the V-groove with weld metal and grind it flush. Use enough current for full penetration of the flange metal. See Figure 5–3 (c).
- Grind the weld flush (on the inside of the flange too if necessary to make it flat) making sure to leave the ground surface as smooth as possible. Any remaining surface imperfections are stress raisers.
- Cut a ½ × 1½-inch reinforcement bar from mild carbon steel 12 to 15 inches (300 to 380 mm) long. Center it on the crack.
- Using SMAW with a low-hydrogen electrode weld this bar to the middle of the flange. The 1½ inch dimension of the bar is vertical. See Figure 5–3 (d).
- Using a grinder, gouge a bevel groove through the unwelded side of the reinforcement bar to sound weld metal on the other side, Figure 5–3 (d).

- Place a fillet weld in the gouged groove using SMAW with a low-hydrogen to secure the other side of the reinforcement bar. See Figure 5–10 (e).
- Do *not* make welds perpendicular to the length of the channel at the ends of the reinforcement bar.

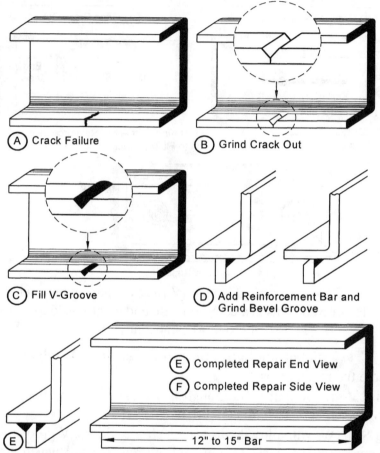

Figure 5–3. Case II: Crack on bottom flange perpendicular to web only.

Case III—Crack on bottom flange perpendicular to web and extending up into web.

This is what happens when Case II is left unrepaired. See Figure 5–4.
The repair steps are:
- Grind out crack with a V-groove halfway through the thickness of the flange along the path of the crack.
- Drill a 3/8- to 1/2-inch (10 to 13 mm) diameter hole at the end of the crack in the web to relieve stress.
- Use reinforcement methods of Cases I and II to add both a bottom reinforcement bar and a reinforcement plate inside the channel web.

WELDING FABRICATION & REPAIR   145

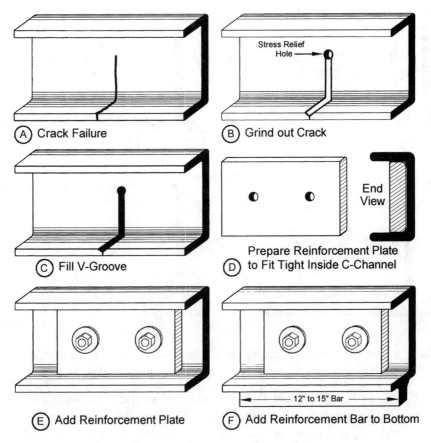

Figure 5–4. Case III: Crack begins on bottom flange and extends into web of channel.

Is there another approach to repairing cracked C-channels?

Another and more traditional C-channel repair approach is to weld reinforcement plates to the web instead of bolting them, or to weld a reinforcement bar along the bottom flange, but this is less commonly done today. If you choose welding instead of bolting for C-channel web repair, be sure to place welds *parallel* to the channel. Do not place any welds perpendicular to the channel. Welds perpendicular to the channel concentrate stresses on just one section of the weld. This is because end welds on the patch plate prevent beam stress from being distributed evenly along the weld length. They become a new stress raiser and will produce near-term failure. See Figure 5–5.

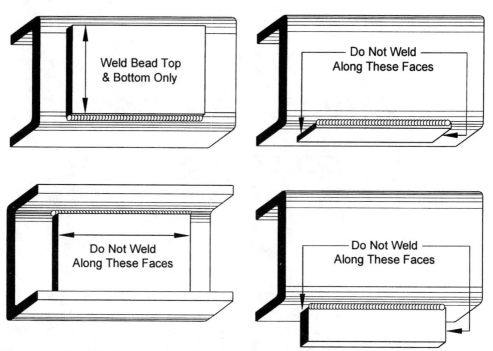

Figure 5–5. Welding reinforcement plates to C-channels: Welds parallel to C-channels are acceptable; perpendicular welds are not.

Caution: Remember that these failures occurred in the first place because the member was stressed beyond its design capacity by excessive loads and fatigue. Dump truck body action stresses, hydraulic cylinder loads, road vibration, and truck overloading all contribute to failure. *Failure is likely to happen again, usually in the next weakest location, because the member is subject to the same load conditions that caused the initial failure.*

You have a relatively new model car—one with the newer high-strength steel used to save weight—and have tried an oxyfuel weld repair on a sheet metal body part, but the weld keeps cracking. What is wrong?

This high-strength sheet metal cannot be oxyacetylene welded. It can, however, be successfully GMAW welded. Begin the weld bead on the *outside* edge of the crack and work toward the *inside*, which will keep the inherent weakness of the bead-ending crater away from the metal's edge where it would act as a stress raiser and lead to a new failure. On galvanized body parts, ER70S-3 electrodes should be used as they contain less silicon than ER70S-6 electrodes. Silicone contributes to cracking when mixed with zinc.

What precautions must be observed when mounting a roll bar or safety cage to a unibody vehicle?

Welding Fabrication & Repair

While unibody design provides a strong and rigid vehicle frame, because it is made of relatively light sheet metal, it cannot support the forces that roll cage tubing ends can exert on it. A good solution is to use GMAW and ER708-3 electrode wire to weld the tubing to a mounting plate and have the mounting plate distribute the forces over a wider area than the tubing end alone. Use ER70S-3 electrode wire. Either an all-around fillet weld or several 1/2-inch (13 mm) diameter plug welds are suitable to hold the mounting plate to the unibody box beam, Figure 5–6 (left). When roll cage tubing supports must be mounted to vehicle flooring, a stiff (1/4 inch or 6.5 mm) mounting plate is required to distribute forces over the thinner sheet metal flooring. This may be welded all around or bolted in a sandwich as shown in Figure 5–6 (right).

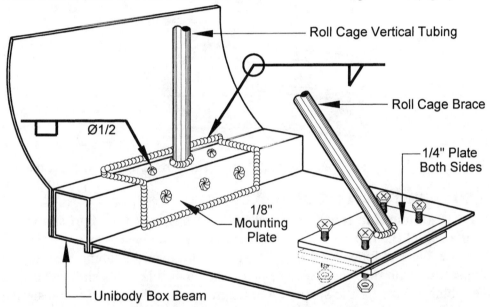

Figure 5–6. Two methods of mounting roll bar tubing on a unibody vehicle: Welding support plate to unibody box beam (left) and bolting mounting plate to sheet metal when welding is not permitted (right).

Preventing Burn-through on Sheet Metal

You have a small hole in vehicle sheet metal and want to patch it with GMAW. Every time you try to make a weld with minimum welder settings, the sheet metal burns through. How can this patch weld be made?

First, wire brush off dirt and any undercoating from the back side of the metal to obtain good heat transfer to the copper heat sink. Wire brush the surface of the work. Then use the heat sink in Figure 5–7 against the back side of the weld when making the weld. The copper heat sink prevents the weld heat from melting too much of the surrounding sheet metal. Finally, grind the weld smooth.

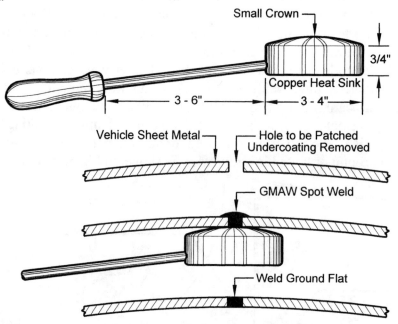

Figure 5–7. Using a copper heat sink to prevent burn-through on sheet metal.

Van Sant Pipe Template Maker

When fishmouthing tubing for roll bars, getting just the right shape cut on the end of the tubing is very difficult. Is there an easy way?

Yes, use Vansant Pipemaster tool. The tool slides over the tubing forming the intersection with the other piece of pipe. The moveable wires slide parallel to the tubing to capture the shape of the fishmouth and this shape is transferred to the tubing needing the fishmouth shape.

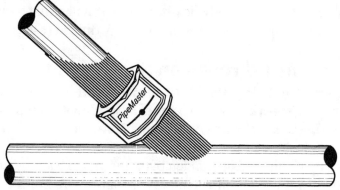

Figure 5–8. Pipemaster tool develops pipe fishmouth shape for any angle and transfers the pattern to workpiece.

Chapter 6

Welding Problems, Solutions & Practices

In every work of genius we recognize our own rejected thoughts.
—Ralph Waldo Emerson

Introduction

There are few genuine welding secrets, but there are many valuable and little known techniques, methods, and procedures that can make any welding job go smoother. Called know-how, they can save time, money, and aggravation. Sometimes they can make the difference between success and failure, profit or loss. Many are based on science, others on experience. This chapter is a collection of this know-how.

Section I – Common Welding Problems & Their Solutions has information to help you solve common welding problems. *Section II – Good Design Practices Checklist* presents a systematic review of key considerations when designing a weldment. *Section III – Controlling Distortion* reviews the cause and practical remedies of weld-induced distortion. *Section IV – Other Design Issues* discusses the sizing of fillet welds and how to avoid fatigue failures. *Section V – Simple Design Ideas* provides insight into common welding applications. *Section VI – Advanced Design Concepts* discusses elastic matching and weld placement. *Section VII – Structural Steel* looks at common materials and practices in that specialty. *Section VIII – Oxyfuel and Lance Cutting* offers problem-solving tips and discusses the applications and capabilities of lances and burning bars. *Section IX – Industrial Fasteners* introduces nuts and bolts you will not find in the local hardware store and how they are used.

Section I – Common Welding Problems & Their Solutions

Extending the Capacity of a Welding Machine

You have two steel parts to join with FCAW or GMAW. One or both are just a little too thick for the maximum available current from your welding machine. How can you get this job done?

Preheat the metal. Use a propane, MAPP® gas, or an acetylene torch to heat the weld area of the two parts up to 300 to 500°F (150 to 260°C), then make the weld. Preheating the metal in the weld area prevents the otherwise cold metal from draining heat away from the weld zone and preventing proper fusion.

Welding Thick to Thin Parts

You need to weld a thin sheet-metal cap onto the end of a much heavier gauge steel tube with GMAW or FCAW. When you go to make the weld you cannot set the current level properly: when it is set low enough to avoid burning through the cap, it will not weld on the heavier gauge tube, and if set high enough to weld on the tube, it burns through the cap. What to do?

There are two solutions:

- Set the welding current so as not to burn through the thinner material, preheat the heavier metal with a torch, then make the weld as if both materials were the same, thinner gauge.
- Set the welding current for the heavier gauge material. When performing the weld, keep the arc on the heavy metal 90% of the time and take short excursions into the thinner material. This takes practice, but works well when mastered.

When welding a thin-wall tube or rectangular shape to a heavier plate, the electrode tends to burn through the thin-walled part. Besides the two methods outlined above is there another solution?

Yes, use a heat sink. If you insert a solid rod inside the tube or a solid bar shape inside the rectangular shape, they will draw heat away from the thin wall and prevent burn-through. In general, solid rod and bar stock is available to fit snugly in most hollow tubes and shapes. Remember to keep the weld away from the tube ends as they are the most vulnerable points of burn-through. See Figure 6–1.

WELDING FABRICATION & REPAIR

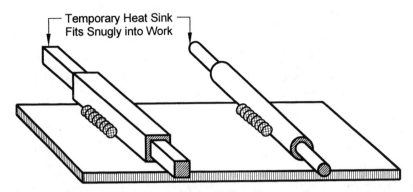

Figure 6–1. Using internal heat sinks to avoid burn-through.

Welding Galvanized or Cadmium Plated Metal

You must weld galvanized or cadmium plated material to another part. How should you proceed?

The best approach is to file or grind off the plating around the area of the weld. The plating not only can contaminate and weaken the weld, but also its fumes are poisonous.

Sealing up Closed Vessels

You are using welding (or brazing or soldering) to seal up a float or close up the ends of a hollow member. The final part of the weld keeps blowing out from the hot air trapped inside the vessel. How can you avoid this problem?

First, drill a 1/16-inch diameter (1.5 mm) pressure-relief hole in the float to vent the weld-heated air to atmosphere and weld all seams shut. Then, seal up the drilled hole with a small weld. See Figure 6–2. Use relief holes when welding structural pipe and tubing with trapped air volume. *Caution: Welding on a closed vessel can be dangerous.* Make sure the interior of the vessel, pipe, or tube is clean and free of flammable or explosive vapors before *starting* to weld.

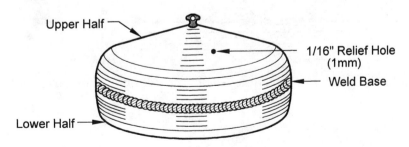

Figure 6–2. Sealing up a float or closed volume.

Welding Screen or Wire Mesh to a Frame

You need to weld screen, wire mesh, or expanded metal to a steel frame with GMAW, FCAW, or GTAW. When you try to make the weld, the screening burns through and the weld is incomplete. What to do?

- Place metal washers over the steel screening (or expanded metal) and clamp washers, screening, and frame together. Vise®-grip Model 9R welding pliers (Figure 1–10) work particularly well for this task. Do not use cadmium plated or galvanized washers; washers should be bare steel. See Figure 6–3 (a).
- As a heat sink, place another larger washer over the washer to be welded in place. This top washer should have a larger hole than the bottom washer so you will avoid welding it in place. Make your plug weld *through* the two holes of the washers and weld only the bottom washer in place. Some tweaking of the welder may be needed to get enough heat to perform the weld, but not so much as to burn away the surrounding screening. See Figure 6–3 (b & c).
- Another approach is to use a metal strip with holes for plug welds at regular intervals in place of the washers. See Figure 6–3 (d).

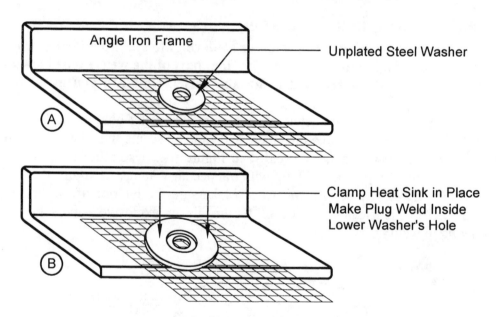

Figure 6–3. Welding screening to a frame. (continued on next page)

WELDING FABRICATION & REPAIR 153

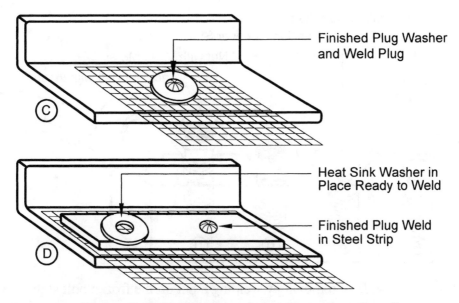

Figure 6–3. Welding screen to a frame. (continued)

Frozen Bolt Removal Methods
What are some ways to remove a seized or rusted bolt in addition to the traditional screw extractor?
Here are two possible methods:
- If the assembly the bolt is in will tolerate heat without damage, use an oxyfuel torch to heat the bolt head and its assembly to just below red heat. A rapid water quench will further help screw removal. Two or more heating and cooling cycles may be needed.
- If the slot, head, or Allen cap screw socket is damaged or missing, place a nut over the bolt head (or remaining stub end), hold this nut in place, and fill the inside of the nut with weld metal using any welding method. This weld will join the nut to the bolt stub. Then put a wrench or socket on the nut and back out the bolt. This has the advantage of simultaneously providing a new gripping point *and* applying heat to the fastener. See Figure 6–4.

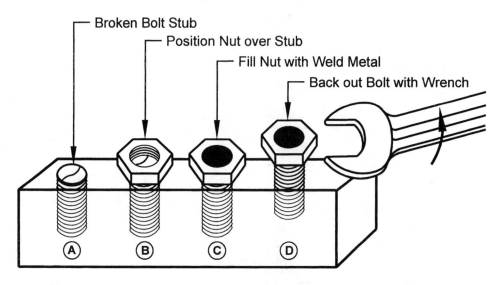

Figure 6–4. Bolt removal by welding a nut on the frozen bolt stub.

Building Up a Worn Shaft

You have a badly worn shaft. What is the best way to build it up with weld metal?

FCAW, GMAW, or GTAW beads can be used. There are four points to remember to get satisfactory results:

- Place weld beads *parallel* with the axis of the shaft.
- Put one bead down, rotate the shaft 180°, then place the next weld bead. This will balance weld stress forces and virtually eliminate heat-induced distortion. Placing sequential beads next to one another will cause the shaft to curve into a banana shape. This process is best done on welding rollers to make handling the shaft easy.
- Weld beads must be placed tightly *onto* one another with 30 to 50% overlap to provide a smooth surface after final machining.
- SMAW and FCAW will require thorough flux removal by brushing and chipping between each bead.

See Figure 6–5.

WELDING FABRICATION & REPAIR

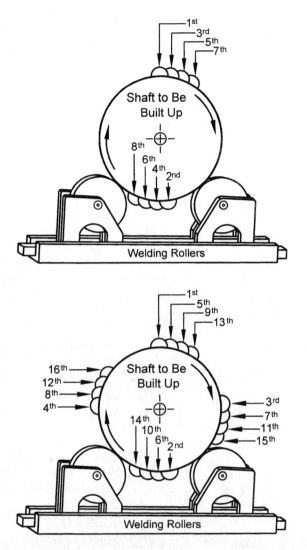

Figure 6–5. Building up a worn shaft with weld metal.

It is also possible to add a weld bead every 90 degrees (at the 3, 6, 9, and 12 o'clock positions) to further reduce distortion effects. In some applications, as on bronze or copper parts, adding braze metal may be better than welding.

Removing a Stuck Steel Bearing

You have a steel bearing stuck in place that does not want to come out. How can welding be used to remove it?

Place a weld bead on the circumference of the *inside* face of the stuck bearing. The combination of the tension of the weld bead reducing the bearing diameter and the heat of the welding process should free the bearing.

A 4-inch diameter steel pipe will shrink about 0.050 inches across its diameter if the bead runs completely around its circumference, Figure 6–6.

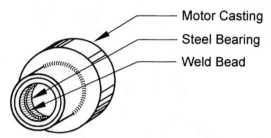

Figure 6–6. Removing a stuck bearing with a weld bead.

Stopping Crack Propagation in a Plate

A tank or ship plate has developed a crack. How can it be stopped?
Drill a hole at the end of the crack to distribute the end stress over a wider area and then put down a trapezoidal padding to add strength to the plate where the crack is headed. See Figure 6–7.

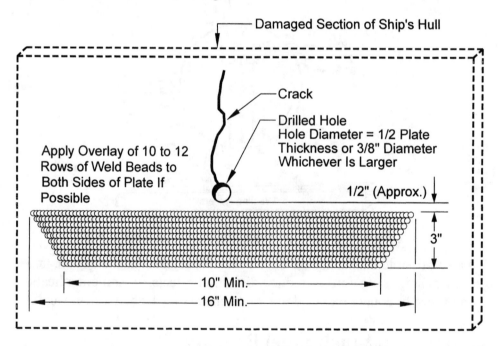

Figure 6–7. How to stop a crack in a plate.

Securing Welded Pads for Machining

Frequently a steel pad is welded to the surface of a steel plate. The fillet welds around the outer edge of the pad tend to force the center of the pad to lift and pull away from the steel plate below it from angular distortion, Figure 6–8 (upper). This separation may complicate machining and threading operations. How can this problem be solved?

Use plug or slot welds within the area of the pad to pull it back into place against the main panel, Figure 6–8 (lower).

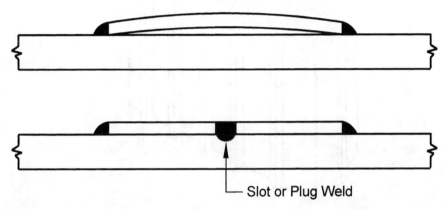

Figure 6–8. Using plug or slot welds to pull a steel pad against its base plate.

Adding Thickness

Sometimes an increased panel thickness is required in a small area and not over the entire panel. How can this be done?

Insert a thicker piece of metal into the panel where needed and secure it by welding, Figure 6–9. Such an insert will provide the thickness to perform subsequent machining, boring, or drilling and can often replace a much heavier one-piece panel or a casting.

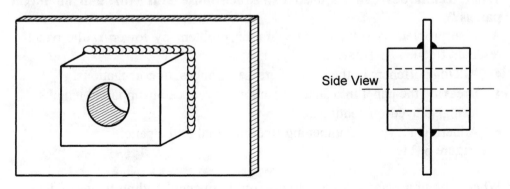

Figure 6–9. Inserting a thick plate into a panel.

Stiffening Flat Panels
What is a standard way to stiffen panels to resist loads?
Weld angle iron on the panels vertically, Figure 6–10.

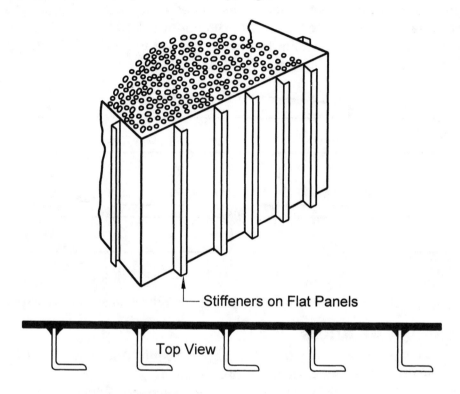

Figure 6–10. Adding angle iron to stiffen panels.

Controlling Noise and Vibration
What techniques can be used to reduce noise and vibration on metal panels?
A noise problem is solved as a vibration problem by lowering the panel's resonant frequency. This is done by:
- Adding stiffeners in the form of creases, flanges, or corrugations.
- Breaking the panel into smaller sections to reduce unsupported lengths.
- Using sprayed-on coatings.
- Adhering a layer of dampening fiber material to the panel.

See Figure 6–11.

When vibration occurs at a relatively low frequency, adding metal stiffeners, as in Figure 6–12, often works.

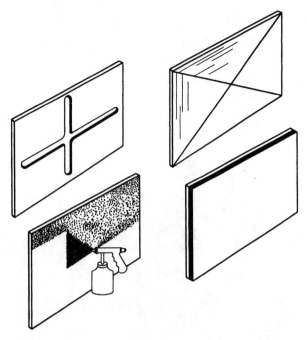

Figure 6–11. Four ways to reduce noise by increasing the resonant frequency of the panel.

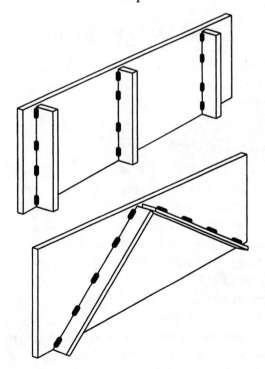

Figure 6–12. Adding stiffeners to reduce low-frequency vibration in panels.

Clamping Work at Right Angles

You need to hold one plate at right angles to another to make a fillet weld. You have only C-clamps available. How can this be done?

Use a block of steel, or even a brick if necessary, and set up the weld as shown in Figure 6–13.

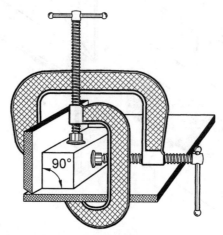

Figure 6–13. Holding work for a fillet weld with two C-clamps and a rectangular block.

Section II – Good Design Practices Checklist

This Section contains a series of design suggestions; they point out potential problems, ways to save time and money, and considerations in designing a weldment and planning its manufacture.

Analysis of the Design

- The design should satisfy the part's strength and stiffness requirements, but must not be overdesigned. Have engineers check the part for safety. They may discover that the stiffness requirements are set too high, causing overdesign. This costs money in extra material, welding, handling, and shipping. Overdesign may also impose added long-term costs on the end-user in fuel, energy, and maintenance. Be sure to have engineers review the new design for safety.
- Specify the appearance required on welds to avoid unnecessary grinding and finishing. Appearance for its own sake usually increases cost more than necessary. Many welds are completely hidden from view. The weldor is not likely to know which welds are critical appearance-wise and which are not unless the print specifies them.

WELDING FABRICATION & REPAIR

- If the product must be fabricated to a code, check restrictions to determine that the most economical method now allowed is used.
- Use deep sections to resist bending.
- Symmetrical sections are more efficient for bending resistance.
- Weld the ends of beams to rigid supports. This increases strength and stiffness, Figure 6–14.

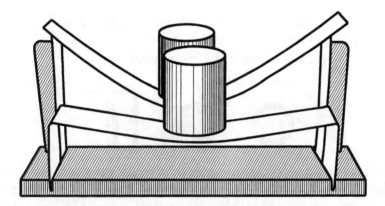

Figure 6–14. Each beam is the same material, width, thickness, and carries the same load. The beam fixed at its ends deflects less than the unfixed beam.

- Use closed sections or diagonal bracing for torsion resistance. A closed section may be several times better than an open one, Table 6–1. Proper use of stiffeners will provide rigidity with less weight, Figure 6–15 through 6–17. Diagonal stiffeners as in Figure 6–17 can often provide the rigidity to replace a heavier casting with a weldment for a machine base.

	Angle of Twist				
	A	B	C	D	E
All Loadings Identical	t = .060 3½	t = .060 2⅛ ¾	t = .060 1.15	t = .060	t = .060
Actual Twist	9°	9.5°	11°	Too Small to Measure	Too Small to Measure

Table 6–1. Five steel shapes and their measured twist when subjected to the same torsional load.

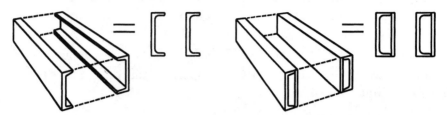

Figure 6–15. Torsional resistance of a frame is approximately equal to the sum of the resistance of its individual members. Closing the C-channels greatly increases their torsional resistance.

Figure 6–16. Round sections are better than rectangular sections in resisting torsional loads because shear stresses are uniformly distributed around the circumference, leaving no stress concentrations; round sections also have equal bending resistance in all directions.

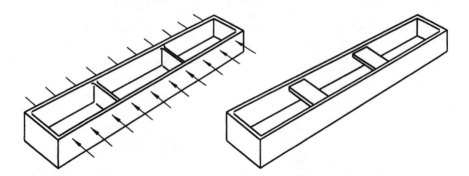

Figure 6–17. Vertical stiffeners (left) are not as effective in resisting compressive loads as horizontal stiffeners (right). Vertical stiffeners must be used in castings because of coring constraints, while horizontal stiffeners can be used in a welded design.

- Diagonal bracing is much more effective than longitudinal vertical stiffeners in resisting torsion. Figure 6–18 shows two steel machine bases. The base on the left is made of 1-inch plate and the one on the right of 3/8-inch plate. They have about equal resistance to twisting, but the diagonally reinforced design offers a 60% weight savings, a 78% reduction in welding, and a total manufacturing cost savings of 54% over the longitudinally reinforced design.

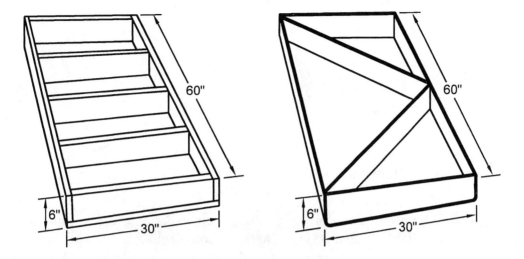

Figure 6–18. Diagonal bracing increases torsional rigidity much more than straight-across bracing.

- Specify non-premium grades of steel wherever possible. Remember that higher carbon and alloy steels require preheating and frequently postheating, which add cost. Try to use higher grades of steel only where required and use mild carbon steel in the rest of the structure.
- Remember that high-strength steels and other premium materials are not available from stock in as wide a range of standard mill shapes as the lower-priced mild steels.
- If only the surface properties like wear-resistance of a higher-priced or difficult-to-weld material are needed, consider using a mild steel base and hardsurfacing to obtain the desired properties.
- For maximum economy and minimum delivery time, use plate, bar, and shapes that are standard.
- If a bar or plate must be machined, ground, or hardsurfaced, dimension the section so the initial plate and bar sizes can be readily obtained from plant or vendor inventory.
- Be sure to provide maintenance accessibility. Do not bury a bearing support, or other critical wear point within a closed-box weldment. The same applies to electrical and hydraulic lines and components.
- Sometimes sections can be designed round so that automatic welding can be used more advantageously, Figure 6–19.

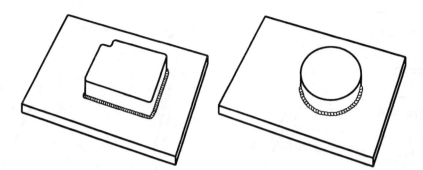

Figure 6–19. Design the weld shape straight or circular to accommodate the automatic welding.

- Check with the shop for ideas where their experience can contribute to better methods or cost savings. Do this before firming the design.
- Check tolerances and press fits first specified. The shop may not be able to hold the specifications economically. Also, close tolerance fits may not be necessary.

Layout

What are the considerations for parts layout?
They are:

- Layout for the fewest number of pieces. This will reduce assembly time and the amount of welding, Figure 6–20.

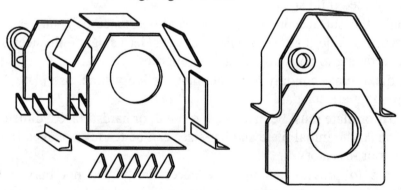

Figure 6–20. Initial design with 21 parts (left) and the improved design with three parts (right).

- Consider layout and design alternatives as they may save materials and welding time. The material cost, cutting cost, welding cost, and the alternative uses of the cutout (excess or scrap) material must be considered before deciding between the approaches of Figure 6–21 and Figure 6–22.

WELDING FABRICATION & REPAIR

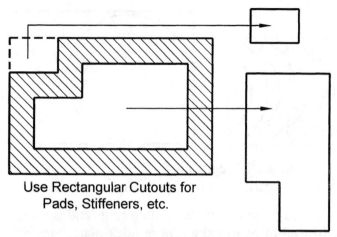

Figure 6–21. If there was an immediate use for the cutouts from this frame, this cut-out approach might make more economic sense than the splicing approach of Figure 6–22.

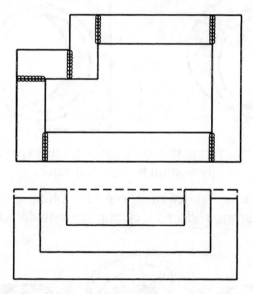

Figure 6–22. Imaginative alternatives can save materials. This frame is made from pieces instead of cut out of a much larger plate.

- When fabricating rings, remember that they can be cut from a single plate, or they can be welded from nested sections on a plate. As in the above example, part dimensional tolerances, material cost, cutting cost, weld cost, and the alternative uses of the cutout material must be calculated before the best approach is known. If material delivery is a problem, the cut segments might get the job done without having to wait for more plate. See Figure 6–23.

Figure 6–23. Cut nested segments for a ring from thick plate to reduce material cost or improve delivery time.

- When tolerances permit, consider rolling rings and welding them up from bar stock instead of cutting them from thick plate. Material waste will be reduced, Figure 6–24.

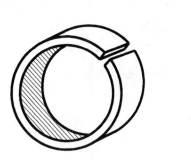 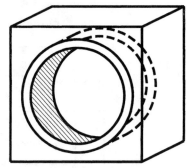

Figure 6–24. Roll rings from bar stock and weld them shut instead of cutting them from heavy plate stock.

- If a number of rings are required, consider rolling a plate into a cylinder, welding up its seam, and flame cutting the cylinder into rings, Figure 6–25.

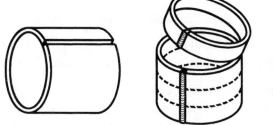 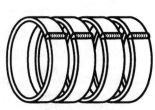

Figure 6–25. Making rings from plate by rolling, welding, and cutting.

- Very complex parts can be fabricated by welding components together. This can save weight, materials, and machining time, Figure 6–26.

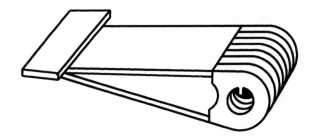

Figure 6–26. Complex made part by assembling cut parts and welding them.

- Putting a flange on a flat plate increases its stiffness at little cost, Figure 6–27.

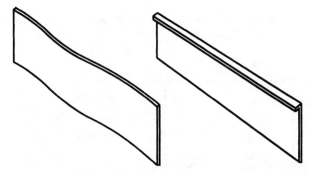

Figure 6–27. Adding a flange to increase a plate's stiffness.

- Bending up the edge of one sheet before welding it onto the next provides a low-cost stiffener, Figure 6–28.

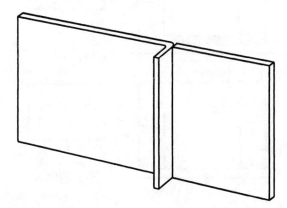

Figure 6–28. Flange between welded panels provides a stiffener.

- Consider using a corrugated sheet for extra stiffness, or press indentations into a plate to add stiffness, Figure 6–29.

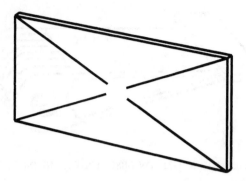

Figure 6–29. Indentations in a flat panel increase its stiffness at little cost.

- Review the design carefully to see if it can be improved to save materials and welding without affecting the strength of the end product, Figure 6–30.

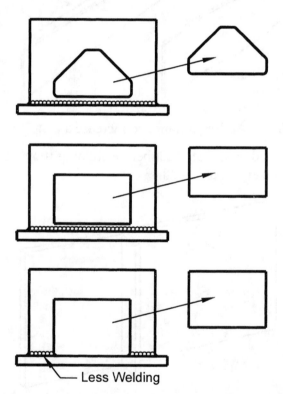

Figure 6–30. Example of design changes saving both welding and materials.

- Check to see that the weld seam location is in the optimum location based on the welding and fabrication process, Figure 6–31.

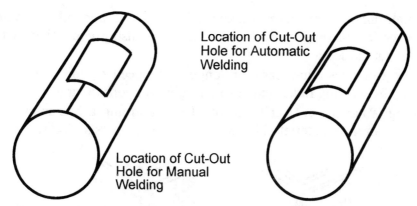

Figure 6–31. Moving a weld seam can avoid wasted welding or make the use of automatic welding equipment possible.

Plate Preparation

- When choosing the best method of producing weldment blanks from plate the choices are:
  - Flame cutting
  - Shearing
  - Sawing
  - Punch press blanking
  - Nibbling
  - Lathe cut-off for bar and tube stock
- When considering the method of producing weldment blanks, remember to factor in the accuracy and edge quality obtainable. Also, remember to factor in the extra stock needed to allow for later application of edge or groove preparation.
- When proposing to combine the cutting of a blank to size and preparation of its edge for welding, remember that not all welds are continuous. A continuously beveled edge that is not continuously welded may be undesirable on exposed joints.
- For a single-bevel or single-V plate preparation, use a single tip flame-cutting torch; conversely, for a double-bevel or double-V plate preparation use a multiple-tip flame-cutting torch so this can be done in one pass of the cutting machine.
- A thick plate is sometimes prepared with a J- or U-groove as it requires less weld metal than a double V-groove.
- Consider whether a steel casting or forging can eliminate a complicated section of the weldment and will simplify the design problem and cost of manufacture.

- A small amount of hardsurfacing alloy can be applied by welding wherever it will do the most good instead of using expensive material throughout the section.
- Wherever flanges, lips, ears, or tongues are needed, consider building them up by welding rather than using forgings or considerable machining.
- Welding is not the answer to all problems. Forming a corner instead of welding can save material, fabricating, and welding costs, Figure 6–32.

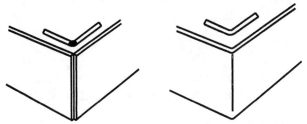

Figure 6–32. Forming a corner can be more economical than welding one.

Welded Joint Design

- Avoid joints that create extremely deep grooves. The joint formed by the meeting of a round or tubing with a flat surface of another round or tubing is one example. This presents two procedure problems: getting proper fusion to the root and not burning through the thin wall while filling he joint with weld metal, Figure 6–33.

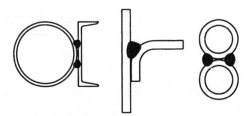

Figure 6–33. Avoid joints like these to avoid poor fusion and burn-through problems.

- Use minimum root openings and included angles in order to reduce filler material required, Figure 6–34.

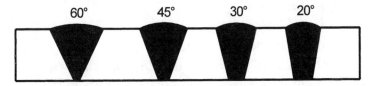

Figure 6–34. Minimum root openings and included angles minimize welding metal and time.

- On thick plate use a double–V instead of single–V joint to reduce the amount of weld metal, Figure 6–35.

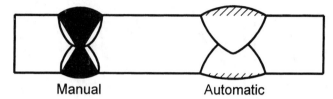

Figure 6–35. Use double-V instead of single-V on thick plate.

- Sometimes a single joint can be used to join three parts at once, Figure 6–36.

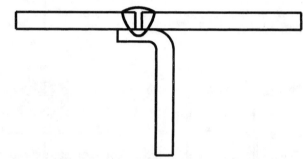

Figure 6–36. Joining three parts with one weld. Gap between top two plates allows for better fusion among all three parts.

- Check to see that the joint location allows access for making the weld, Figure 6–37.

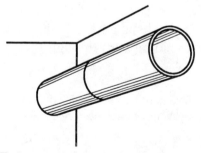

Try to avoid placing pipe joints near wall so that one or two sides are inaccessible. These welds must be made with bent electrodes and mirror.

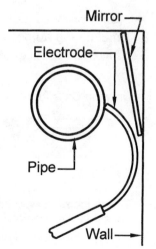

Figure 6–37. Designing for weld access. (continued on next page)

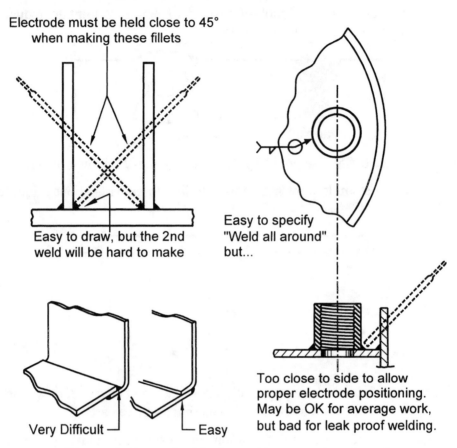

Figure 6–37. Designing for weld access. (continued)

- Choose weld designs that have adequate strength, use minimum weld metal, and do not present burn-through problems for the weldor, Figure 6–38. See Figure 6–39 for guidelines on avoiding burn-through.

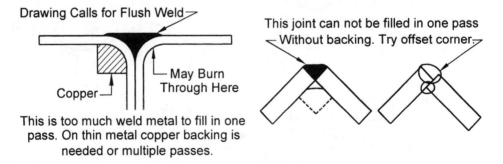

Figure 6–38. Choose weld designs to avoid burn-through. (continued on next page)

Welding Fabrication & Repair

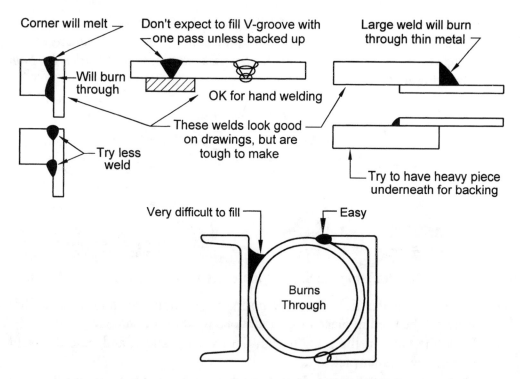

Figure 6–38. Choose weld designs to avoid burn-through. (continued)

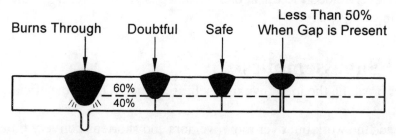

About 60% penetration is all that can be safely obtained with one pass wthout backing on a joint with no gap, even less when gap is present

Figure 6–39. How to avoid burn-through.

Weld Size & Amount

- Be sure to use the proper amount of welding—not too much and not too little. Excessive weld size is costly.
- Specify by print or *Standard Shop Practice* that only the needed amount of weld should be furnished. The weld specified by the designer has a built-in safety factor. Don't add still another safety factor.

- Remember that the leg size of fillet welds is especially important, since the area or amount of weld filler required increases as the square of the increase in leg size. *Doubling the leg size increases weld filler by four times,* Figure 6–40.

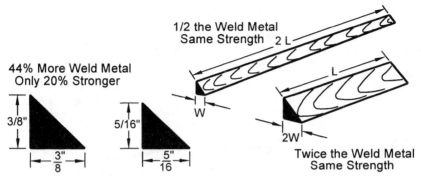

Figure 6–40. Doubling the leg size increases weld filler metal four times.

- Sometimes under light loads or no-load conditions, an intermittent fillet weld can be used in place of a continuous weld of the same leg size.
- Place the weld in the section with the least thickness, and base the weld size on the thinner plate.
- Stiffeners or diaphragms do not need much welding; therefore they are often overwelded. Reduce the weld leg size or length of the weld if possible.

Use of Subassemblies

Consider breaking the job down into subassemblies. This is beneficial because:
- Spreads the work out over more weldors and shortens delivery time.
- Provides better welding access.
- Reduces the possibility of distortion.
- Makes machining of small areas more convenient.
- Facilitates stress relief in limited areas.
- Allows leak testing of compartments and chambers.
- Permits in-process inspection before job has progressed too far to rectify errors.

Assembly

- Clean work of oil, rust, and dirt *before* welding.
- Check fit-up. Gaps are costly to remedy.
- Clamp work into position and hold during welding.

Welding Fabrication & Repair

- Use jigs and fixtures to hold parts with proper fit-up and to maintain alignment during welding.
- Preset joints to offset expected contraction.
- Prebend the member to offset any expected distortion.
- Use strongbacks.
- Where possible, break the weldment into natural sections so the welding of each can be balanced about its own neutral axis.
- Weld the more flexible sections together first so they may be more easily straightened before final assembly of the member.

Welding Procedure

- Try to improve operating factor; use weldor helpers, good fixtures, and handling equipment.
- Deposit the greatest amount of filler metal in the shortest possible time.
- Use backup bars to increase speed of welding of the first pass for groove joints.
- Eliminate or reduce preheat by using low-hydrogen electrodes.
- Use manual electrodes down to a 2-inch (50 mm) stub.
- Weld in flat, down hand position if possible. Overhead and vertical welds are more expensive.
- If possible, position fillet welds in the flat (trough) position for highest welding speed.
- With automatic welding equipment, position fillet welds to obtain greater penetration into the root of the joint: Flat plate at an angle of 30° from the horizontal. Vertical plate 60° from the horizontal, Figure 6–41.

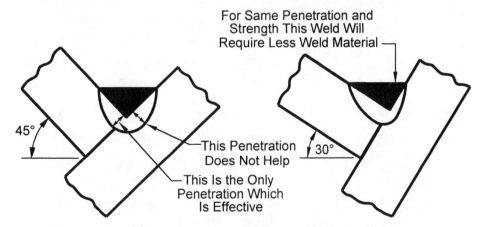

Figure 6–41. Position the joint to minimize weld size without affecting penetration or strength.

- Consider using larger electrodes at higher currents.
- Weld toward the unrestrained portion of the member.
- Use a welding procedure that eliminates arc blow.
- Weld first those joints that may have the greatest contraction as they cool.
- Be sure you are using optimum travel speed and current/voltage settings.
- Use semi-automatic or fully automatic welding where possible and take advantage of its deeper penetration and uniform deposit.

Cleaning & Inspection

- Do not grind the surface of a weld smooth or flush unless required for a reason other than appearance. This is a very costly operation and usually exceeds the cost of welding.
- Reduce cleaning time by using iron powder electrodes on SMAW and GMAW.
- Use antispatter films applied parallel to the weld line.
- Good welds must always be the goal; however, even a "poor" weld is often stronger than the plates being joined.
- Overzealous inspection can run up welding costs rapidly.

Section III – Controlling Distortion

Cause of Distortion

What happens to the physical properties of metals when heated?
Metals' physical properties can change greatly: stiffness (E, or Young's modulus) decreases, the yield point falls and coefficient of expansion increases. Figure 6–42 shows these property changes for mild carbon steel.

WELDING FABRICATION & REPAIR

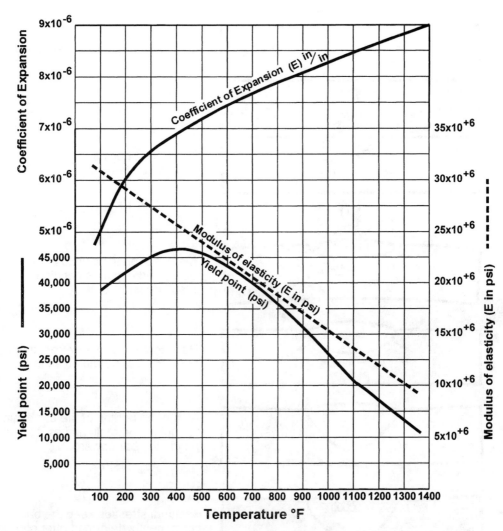

Figure 6–42. How three physical properties of mild carbon steel change at elevated temperatures.

Why does welding cause distortion?

Welding raises the metal along the weld path to very high temperatures. Peak temperatures near the arc are measured in thousands of degrees. This heating causes the metal closest to the weld to expand *more* than the cooler metal more distant from the weld. This cooler metal resists expansion forces from the hotter, weaker metal, causing permanent deformation in the work metal. When the work cools, thermal contraction forces cause metal shrinkage. Some of these contraction forces are dissipated as they distort the work metal. In rigid portions of the work piece, these contractive forces are not dissipated and are stored as residual, or locked-in stresses, Figure 6–43.

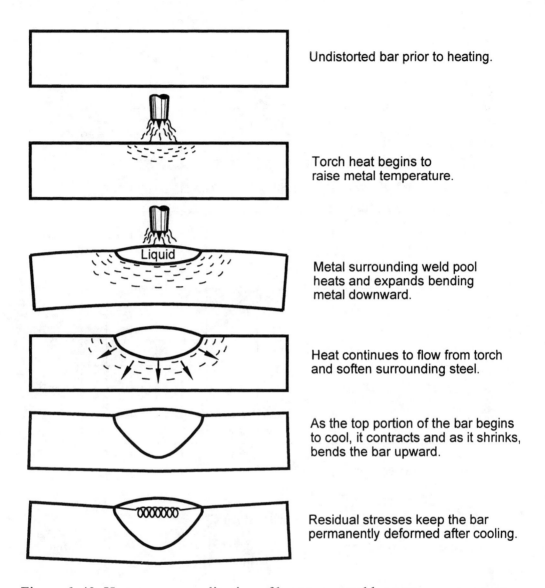

Figure 6–43. How uneven application of heat to a metal bar causes permanent distortion.

How does the application of a weld bead along a steel bar cause the bar to distort?

Shrinkage occurs along the entire length of the weld, Figure 6–44. Shrinkage always occurs in both the applied weld metal and the base metal.

WELDING FABRICATION & REPAIR

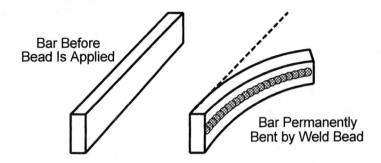

Figure 6–44. How application of a weld bead affects a steel bar.

What factors control the degree of shrinkage and distortion?
- Restraint from external clamping.
- Internal restraint due to the mass of the workpiece.
- Stiffness of the steel itself.
- The amount of heat input and the rate it is applied.
- Cooling rate.

The interaction of these factors is complex. While we can take many practical steps to control shrinkage and distortion, calculating and predicting its effects for all but the simplest welded objects is not possible.

Where can we expect to see these shrinkage and distortion effects?
Longitudinal shrinkage occurs along the axis of the weld and *transverse shrinkage* occurs perpendicular to the axis of the weld, Figure 6–45 (left). If the cross section of the weld has a triangular shape, more shrinkage occurs where the weld is thickest, causing *angular distortion*, Figure 6–45 (right). This shrinkage occurs in both the deposited weld bead metal and the surrounding metal as well.

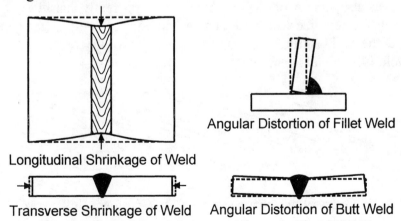

Figure 6–45. Weld shrinkage has several different effects.

The effects of longitudinal shrinkage on the work can be different depending on where the weld is in the structure with respect to the neutral axis, Figure 6–46. See the Strength of Materials Chapter for an explanation of *neutral axis*.

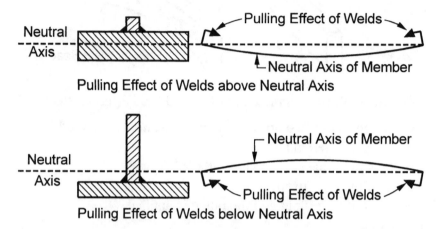

Figure 6–46. Shrinkage effects depend on the weld's location.

Overwelding

What is the largest, single controllable factor leading to distortion?
Overwelding. It is often inadvertently caused by the following chain of events: The designer may specify the next larger weld size because of a lack of confidence in welding. When the part reaches the shop floor, the shop foreman, wishing to play it safe, marks the piece up for the next weld size. The weldor, having just been criticized for making under size welds, makes very sure that these welds are still larger. The result—a 1/4-inch fillet has become a 1/2-inch weld. These people usually do not realize the weld metal increases as the square of the leg size. The apparently harmless 1/4-inch increase in the leg size has increased the amount of weld metal deposited, increased the weld shrinkage, and increased the weld cost by four times. More weld filler metal is not necessarily better, Figure 6–47.

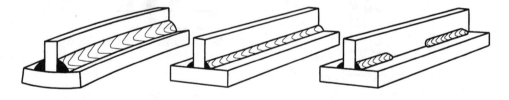

Figure 6–47. Overwelding increases the shrinkage force (left), small leg size decreases distortion forces (center), and decreasing the weld length by using intermittent welds further reduces shrinkage and distortion (right).

Controlling Shrinkage & Distortion

What steps can be taken to reduce distortion?
See Figures 6–48 through 6–50.

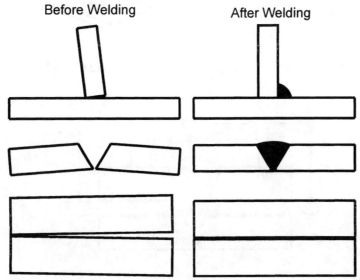

Use pre-setting and let weld shrinkage bring parts into position.

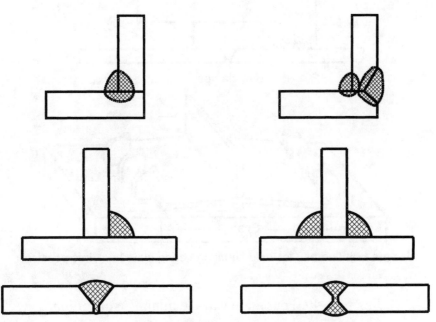

Convert unbalanced joint designs into balanced ones.

Figure 6–48. Two most basic distortion control methods.

182   CHAPTER 6   WELDING PROBLEMS, SOLUTIONS & PRACTICES

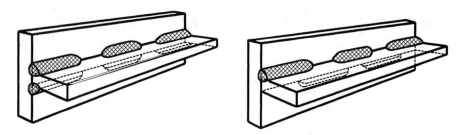

Use chain intermittent and staggered intermittent welds to reduce shrinkage.

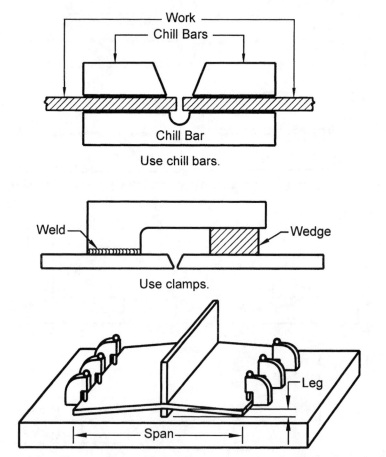

Use chill bars.

Use clamps.

Use pre-bending and let weld shrinkage bring plate back into position.

Figure 6–49. Four more ways to minimize distortion.

WELDING FABRICATION & REPAIR

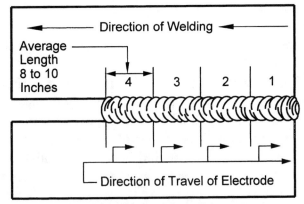

Use back-step welding sequence.

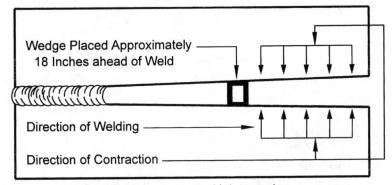

Use wedge to control joint spacing.

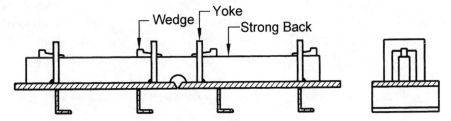

Plates forced into alignment and held by means of strong backs. The pressure being applied by means of a wedge driven in between a yoke and the strong back.

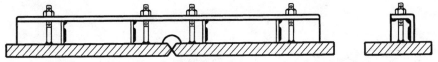

Use strongbacks. Temporary studs to support strongbacks are welded to plate.

Figure 6–50. Three slightly more complex ways to minimize distortion.

Are there additional stress and distortion control methods in use?
Yes, there are several:
- Alternating passes on each side of a multi-pass butt weld to minimize angular distortion.
- Sequence joints in a complex structure to balance shrinkage forces; in general, perform diagonally opposite welds, not adjacent ones. See Chapter 2 – Basic Building Blocks for examples.
- Preheating *may* be useful; it is primarily used to obtain quality welds by reducing the cooling rate, particularly in cold weather. One joint U.S. Navy-U.S. Maritime Administration study did not find it useful in reducing weld distortion.
- Postheating of heavy weldments is often done to relieve residual stresses to prevent *further* distortion from occurring when machining removes additional metal. Removing metal from a structure with residual stress can lead to more distortion.
- Flame straightening, covered in Chapter 4 – Bending & Straightening.

Frequently two or more stress and distortion control steps will be used together to get acceptable results.

Transverse Shrinkage
How does weld size and shape influence weld shrinkage?
Shrinkage is directly related to the amount and shape of metal in the weld. Figure 6–51 (upper) shows that a double-V weld has about half the shrinkage of a single-V weld in the same size plate. Figure 6–51 (lower) shows a direct and linear (straight-line) relationship with the area of the weld for constant plate thickness. In general, we can estimate the transverse shrinkage of a weld to equal 10% of the average width of the weld's cross-section, or

$$\text{Transverse shrinkage} = 0.10 \times (\text{Average weld width})$$

This indicates that minimizing weld size, consistent with the strength required, is desirable to control shrinkage.

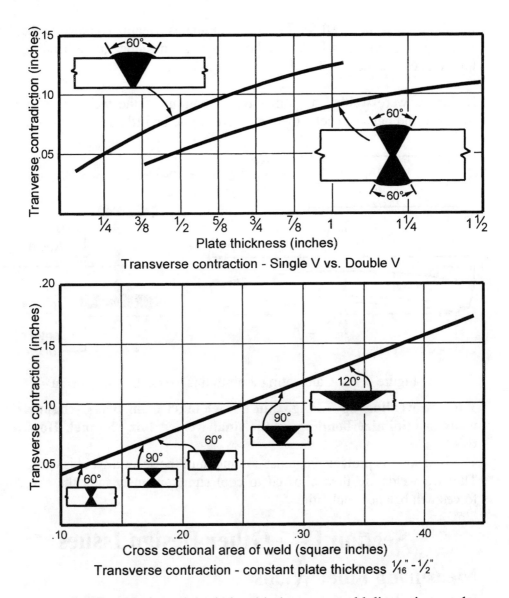

Figure 6–51. Graphs show the relationship between weld dimensions and shrinkage.

Longitudinal Shrinkage in Beams

Can we predict the amount of deflection at the center of an I-beam when a plate is welded to one flange?

Yes. There is excellent agreement between the calculated and actual deflection caused by longitudinal weld shrinkage in a cover plate using the formula in Figure 6–52. The area, A, and the moment of inertia, I, a measure of the beam's stiffness, can be obtained from widely available tables. This formula also works well for T-beams and angles. Beams can also be bent by mechanical bending in hydraulic bending machines and flame bending with torches, see Chapter 4 – Bending & Straightening.

If flanges were welded to both top and bottom of the web, the shrinkage forces would balance each other and there would be little or no deflection.

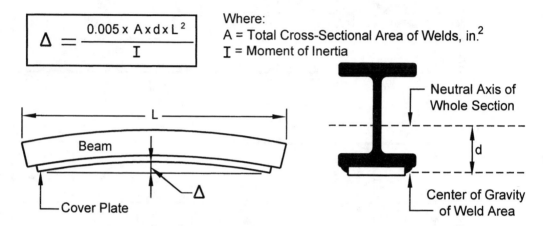

$$\Delta = \frac{0.005 \times A \times d \times L^2}{I}$$

Where:
A = Total Cross-Sectional Area of Welds, in.²
I = Moment of Inertia

Figure 6–52. Calculating longitudinal deflection in a beam.

You are welding up a long, thin box channel from two C-channels and want to minimize bending in the final welded box channel. How to do this?

The first weld is protected against cooling until the second weld is completed. The two welds are then allowed to cool simultaneously, and the shrinkage forces will balance each other.

Section IV – Other Design Issues

Measuring Fillet Welds

How are fillet weld dimensions measured to determine if they meet the drawing's requirements?

Fillet shapes may be concave, convex, or flat. They may have equal legs or unequal ones. However the true fillet size is measured by finding the leg length of the largest isosceles right triangle (a triangle with a 90° corner and legs of equal length) that can be inscribed within the weld cross-section, with the legs in line with the original surface of the metal. The size of a fillet weld is hard to measure without proper gauges. Figure 6–53 shows how to do this.

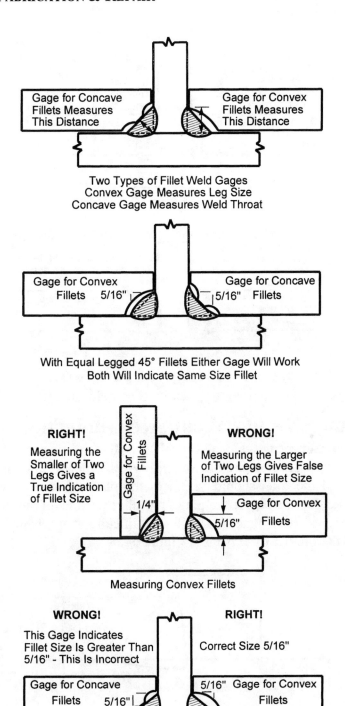

Figure 6–53. How to measure fillet welds. (continued on next page)

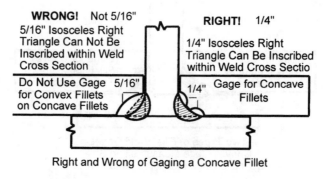

Right and Wrong of Gaging a Concave Fillet

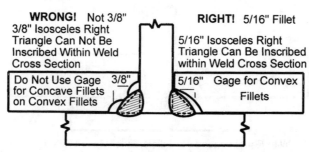

Right and Wrong Method of Gaging a Convex Fillet

Figure 6–53. How to measure fillet welds. (continued)

Minimum Effective Weld Size Guidelines

What is a conservative rule of thumb for determining the size of fillet welds in order to develop the full strength of the plates they join?

The leg size of the fillet should be 3/4 of the plate thickness, assuming the fillet weld is on both sides of the plate, the fillet weld is on the full length of the plate, and if two plates of different thickness are being joined the thickness of the thinner plate is used, Figure 6–54.

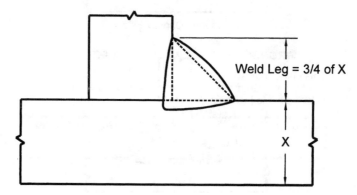

Figure 6–54. Fillet should be 3/4 the thickness of the thinner plate.

What are some recognized guidelines for sizing welds?

The AWS and AISC have used these guidelines in both structures and bridges, Figure 6–55.

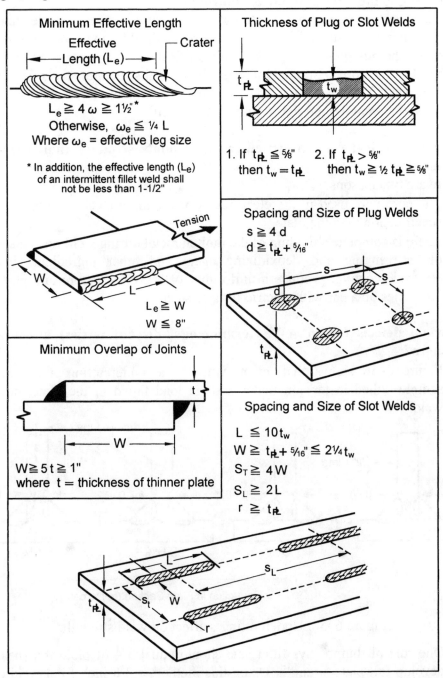

Figure 6–55. AWS/AISC guidelines on minimum effective weld length.

Weld Strength

How strong are typical structural welds between steel plates when made according to industry standards and codes?

Welds have a great reserve of strength over that required by codes and in service. In many cases, this reserve is not recognized by the code bodies. For example, the *minimum* strength of ordinary E60XX electrodes is about 50% higher than the corresponding values of the typical structural steels they join, and, as welded, many of the commercial E60XX steels have about a 75% higher yield strength than the structural steel itself.

Why are typical welded structural steel joints stronger than the base metal alloys?

There are two reasons:
- The core wire used in the electrode is of premium steel, held to a closer specification than the plate.
- There is complete shielding of the molten metal during welding. This, plus the scavenging and deoxidizing agent and other ingredients in the electrode coating, produce a uniformity of crystal structure and physical properties on a par with electric furnace steel.

How do defects (or in AWS terminology, discontinuities) affect weld strength?

- Figure 6–56 shows that, even with severe undercutting, the weakest sample pulled in tension under a static load failed in the plate, not the weld.

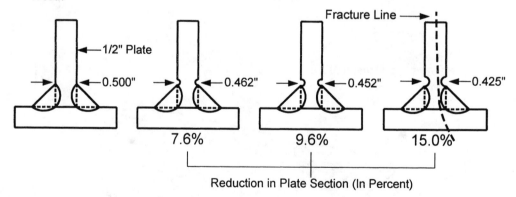

Figure 6–56. Undercutting's affect on weld strength.

- One rule of thumb says fillet size should equal 3/4 of plate thickness to develop full plate strength. Using this method, a 3/8-inch fillet weld on a 1/2-inch plate should be stronger than the plate, but so should 11/32- and

5/16-inch fillets. Not until fillet size was reduced to 1/4 inch did weld failure occur at a stress of 12,300 lbs/lineal inch, more than 5 times the AWS allowable stress, Figure 6–57.

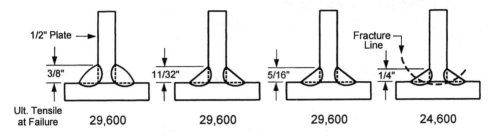

Figure 6–57. Undersized weld size's affect on weld strength.

- Figure 6–58 shows how lack of fusion affects weld strength. All welds were machined flush prior to tensile testing, and weld failure did not occur outside the weld until throat reduction reached 31%.

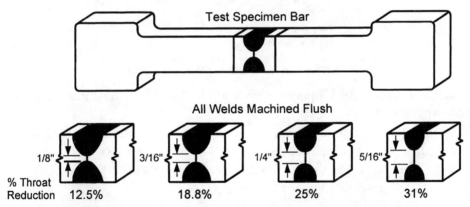

Figure 6–58. Lack of fusion's affect on weld strength.

How much porosity can a weld contain before its strength is affected?
Tests have shown that a weld can have a total void volume of up to 7% before its strength is materially changing its ductility, or tensile and impact strength.

Section V – Simple Design Ideas

Adding Threads and Leveling Jacks
You want to weld a nut concentrically over a hole in a plate or member to provide threads for leveling screws. What is the best way to position the nut over its hole?

- File or grind off zinc or cadmium plating on the nut prior to welding to avoid contaminating the weld with the plating metal.
- Then use another nut and bolt to draw the nut flat and hold it securely against the plate when welding.
- Remove the non-welded nut and bolt when cold.

See Figure 6–59.

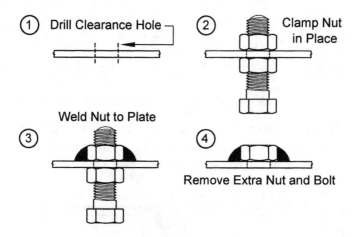

Figure 6–59. Welding a nut over a hole.

What is a quick and inexpensive way to fabricate leveling jacks for a table or frame?

Grind the tips of a hex bolt head down so that the head force fits into the hole of a washer, then weld the washer to the bolt forming a foot. Weld a piece of 1/4-inch (6 mm) rod to the bolt to make turning it easier. Affix a nut to form threads as discussed above. See Figure 6–60.

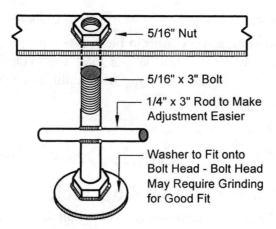

Figure 6–60. Leveling jack.

Pins and Hasps

What is a good way to accurately locate and secure pins or stops on a steel part?
- Locate and center punch the pin position.
- Drill a hole that the pin can slip or be driven into.
- Chamfer the pin end, insert the pin in its hole and make a plug weld on the rear of the pin.
- Plug weld may be ground flush with the back of the member for better appearance.

See Figure 6–61.

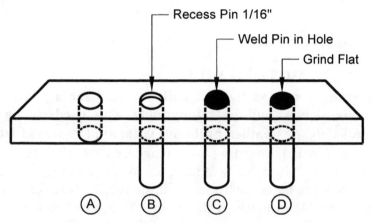

Figure 6–61. Securing a pin with a weld.

How can a hasp or tie point be easily fabricated?
- Cut a chain link with a hacksaw, Sawzall®-type reciprocating saw or Portaband®-type hand-held band saw so the sides of the link are parallel.
- Drill two holes to fit the position and diameter of the cut link sides.
- Insert the link into the holes and secure them in place with plug welds from the rear.

If there is no welding access on the back of the member on which the hasp will be placed, it can be made up onto a small plate and the plate welded to the member. See Figure 6–62.

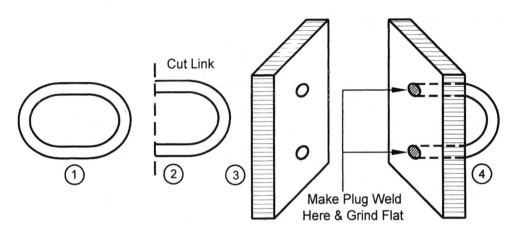

Figure 6–62. Making a hasp or tie point from a chain link.

Closing the Ends of Tubing

You have made a welded item of hollow round tubing or rectangular tubing and must close off the tubing ends. This needs to be done for appearance, safety, or weather-proofing. What are two ways to do this?

- The easiest way is to insert a plastic plug sized for this tubing, Figure 6–63 (top). These plugs are available in round, square, and rectangular shapes. There are two drawbacks to plastic plugs. Most designs and sizes are only available in black. If the product is exposed to the weather, the plastic plugs fail to protect the edges of the tubing from weather; rusting will eventually begin there.
- Another approach is to cut a steel cover for the tubing end and weld it in place. See Figure 6–63 (bottom). This takes more time, but will protect the tubing ends from weather.

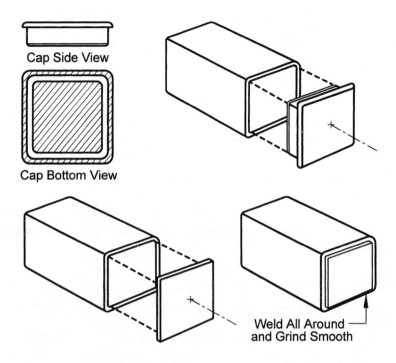

Figure 6–63. Two ways to close off tubing ends: With plastic end plug (top) and with welded plate cover (bottom).

Making Brackets

You need a way to secure an object to a frame member, yet make it easily removable. How can this be done?

Weld a short length of rectangular tubing to the member, then affix the object to a solid square which will slip fit into the tubing. See Figure 6–64. Note that using round tubing and a rod will permit the bracket to swivel, if desired.

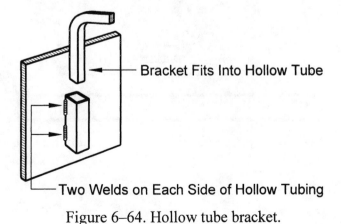

Figure 6–64. Hollow tube bracket.

Securing Hollow Objects
How do you attach a hollow cylinder or hollow casting securely to a metal base, yet make it removable?

- Weld or braze a large-diameter bolt to the plate by its head. Place a weld bead completely around the bolt head.
- It may be necessary to preheat the bolt and plate because of their size, as detailed in Extending the Capacity of a Welding Machine in Section I of this chapter.
- Slide the cylinder or casting over the inverted bolt and secure it with a strongback, washer, and nut.

See Figure 6–65. The inverted bolt provides a large diameter thread to secure the hollow object without cutting threads or drilling a large diameter hole in the mounting plate.

Bolts between 3/8- and 5/8-inch diameter can be readily attached to a 1/2-inch thick steel plate with a 130 A GMAW/FCAW welding machine if torch preheating is used.

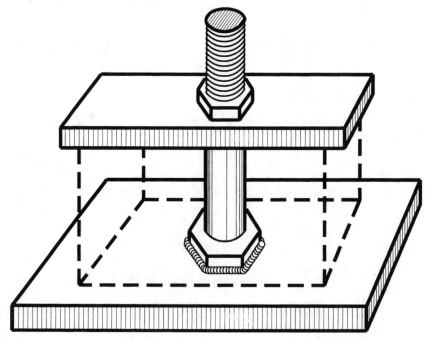

Figure 6–65. Using a welded bolt to secure a hollow object.

Retaining a Nut on a Bolt or All-thread
What are three ways to prevent a nut from being removed from a bolt?
See Figure 6–66.

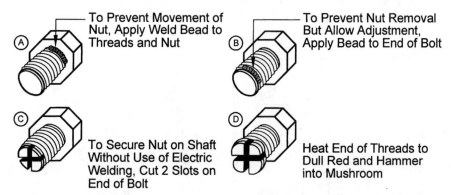

Figure 6–66. Three ways to retain a nut: (a) Weld nut onto thread, (b) Apply weld bead on end of threads, (c) Make hacksaw cuts in end of shaft, then (d) apply torch heat and mushroom the bolt end to prevent nut removal.

Adding Threads to a Tube or Pipe
Most tubing walls are too thin to accept threads. How can threads be added to tubing?
A threaded insert can be placed inside the end of the tubing and secured in place with one or more plug welds. This is a common practice in automotive suspensions and aircraft control rods. See Figure 6–67.

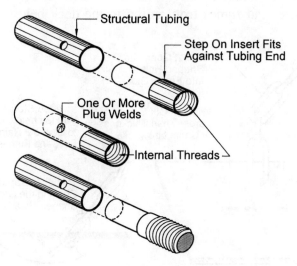

Figure 6–67. Adding threads to tubing with inserts and plug welds.

Making Cleats in a Plate

How can cleats be applied to a steel plate without welding?

Cut the pattern shown in Figure 6–68 to form a cleat. The sunken Russian submarine *Kursk* was lifted from the ocean bed using cleats like this cut into her hull with a plasma torch. Twenty-six steel cables in two rows made the lift.

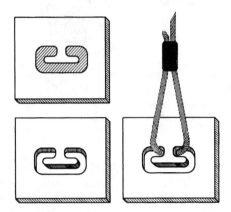

Figure 6–68. Cleat cut into steel plate.

Deck Tie-Downs for Chain

If you have a chain instead of a cable and want to secure it to a deck or bulkhead, how can one be installed without welding?

On shipboard, securing loads to decks is a common problem. Flame-cut cross-shaped holes in the deck provide a secure, convenient means to attach tie-down chains to decks. A steel pan is welded below the tie-down to prevent water, fire, and fumes from going to the deck below. See Figure 6–69.

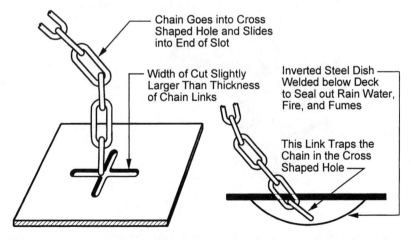

Figure 6–69. Deck tie-down for chain.

Design for Top-Mounted Bearing Supports
What are some of the ways to design a top-mounted bearing support from welded components?
There are many possibilities, Figure 6–70.

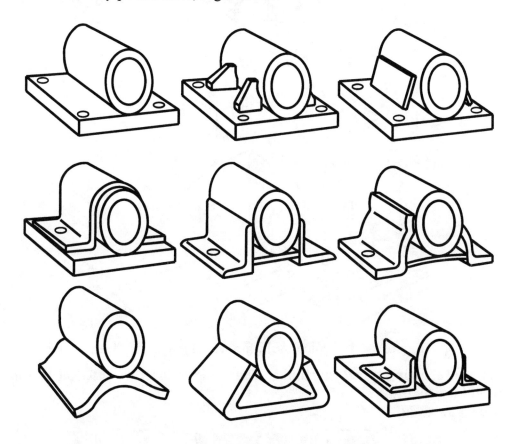

Figure 6–70. Ways to design bearing supports with weldments.

Designs for Machine Bases
What are some of the ways to design machine feet from welded components?
See Figure 6–71.

200 CHAPTER 6   WELDING PROBLEMS, SOLUTIONS & PRACTICES

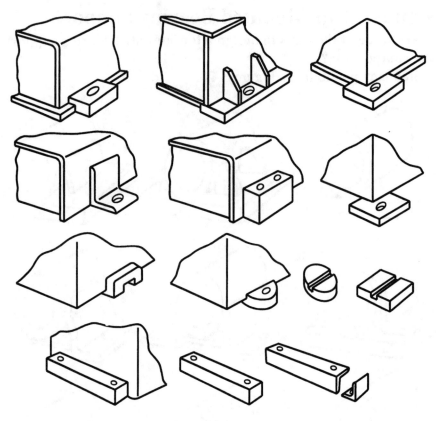

Figure 6–71. Typical weld-fabricated machinery feet.

What are some typical machinery feet designs that are integral to the machinery housing?
See Figure 6–72.

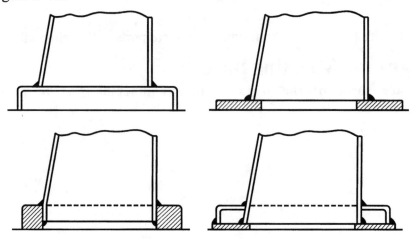

Figure 6–72. Typical feet designs for heavy machinery.

Making Welded Brackets to Replace Castings

What are two designs that allow fabrication of brackets to replace castings?

Figure 6–73 shows two ways a strong bracket may be fabricated by welding.

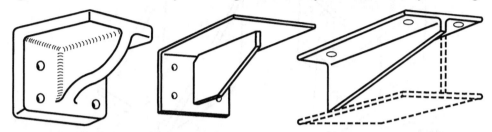

Figure 6–73. The casting to be replaced (left), a two- or three-piece design (center) and another design using cutting operations but no welding (right).

Figure 6–74 shows the evolution from a casting, to a three-piece design, to a final two-piece design.

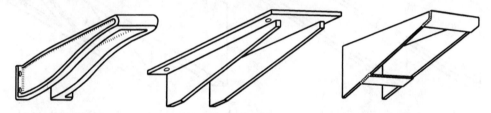

Figure 6–74 shows the design evolution from a casting to a lighter and less costly bracket design. Another advantage of welded brackets over cast ones is that turn-around and delivery time is usually much shorter.

Figure 6–75 shows the design progression from a casting to a weldment and its advantages.

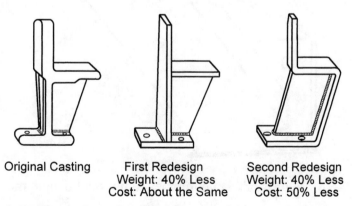

Original Casting

First Redesign
Weight: 40% Less
Cost: About the Same

Second Redesign
Weight: 40% Less
Cost: 50% Less

Figure 6–75. Evolution of welded bracket design to replace a casting.

Figure 6–76 shows the advantages of combining bending and welding processes to develop a bracket.

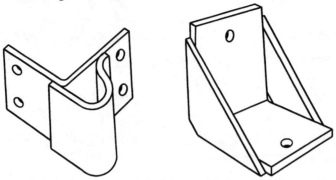

Figure 6–76. Brackets made by combining bending and welding.

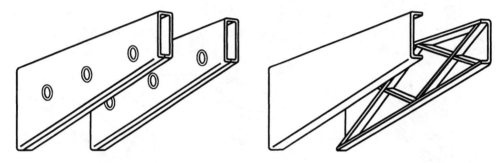

Figure 6–77. Brackets for heavy loads with internal stiffeners.

Figure 6–78 shows how combining a section of rectangular tubing and a right angle bracket with welding can produce an attractive heavy-duty bracket.

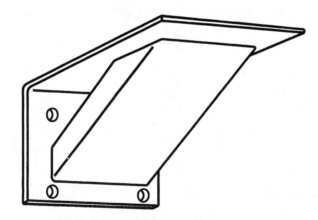

Figure 6–78. Welded bracket combines tubing section and metal angle.

Section VI – Advanced Design Concepts
Elastic Matching
The roller in Figure 6–79 deflects under load, causing misalignment and severe wear in its bearings. How can this problem be eliminated without putting the bearing on a swivel?

The solution is to design a welded steel bearing support that deflects the same angle under the same load that the shaft does, making the shaft and bearing stay in alignment. This technique is called elastic matching, Figure 6–79.

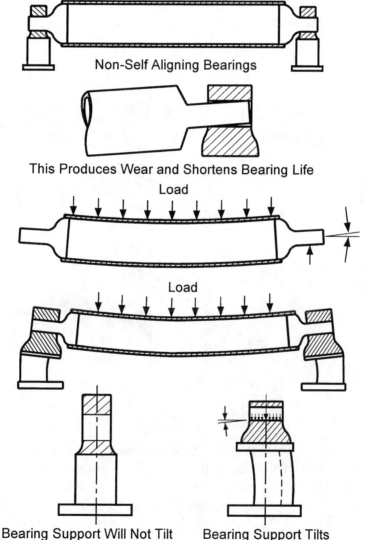

Figure 6–79. Elastic matching.

Locating Welds at Zero Stress Points

Figure 6–80 (a) shows a cross-supporting beam that must have minimum deflection *and* be removable. One way to do this is to weld the I-beam to plates and bolt the end plates to the structure. While this will work, it may be difficult to remove, depending on the rigidity of the surrounding structure. Is there a better way to do this?

Yes, Figure 6–80 (c) shows the bending moment (or force) on a beam with a concentrated load at its center and fixed ends. If we splice the beam at these two zero-stress points where there is no bending force on them, the connection will only have to support the vertical shear load. A simple welded clip on the bottom of the end beam stubs and a bolt in the middle of the beam web will support the load as rigidly as Figure 6–80 (b) design, yet be easier to remove.

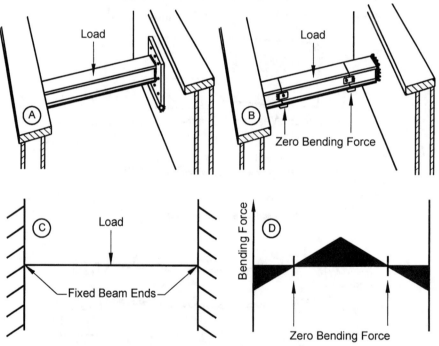

Figure 6–80. Placing welds at zero-stress points.

The vertical welds on the angle iron assembly shown in Figure 6–81 (upper) keep failing when subjected to the indicated cyclic load. What is going on and how can this problem be fixed?

As the cyclic load is applied, the steel angle flexes and puts the two vertical welds into horizontal shear pulling against each other. These welds carry the horizontal shear loads from the stretching of the beam *in addition* to the

WELDING FABRICATION & REPAIR

vertical shear load. If we replace these two vertical welds with a single horizontal weld, these horizontal shearing forces will be reduced, Figure 6–81 (lower). Also, the closer the weld is to the thicker part of the angle iron, in this case the top section, the smaller the stress on the weld. This is because the additional metal in the top of the angle reduces the amount of stretching in the angle iron.

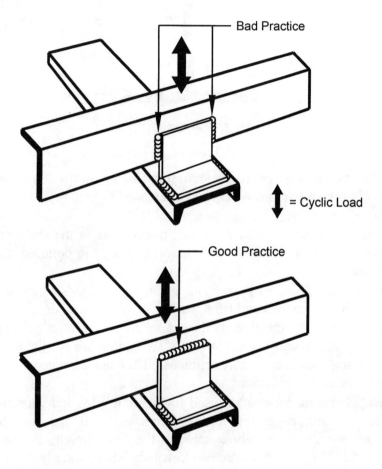

Figure 6–81. Replacing two welds in shear with a single weld along the neutral axis.

Section VII – Structural Steel

I-Beams
What are the parts of an I-beam and what are its key dimensions?
See Figure 6–82.

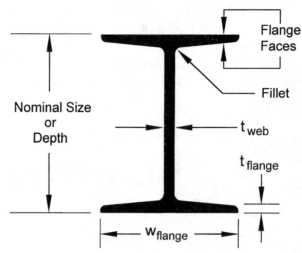

Figure 6–82. Parts of an I-beam.

What are the three most common structural beam shapes, what are the differences among them, and where are they used?

They are the:

- S-beam or *American Standard Shape* I-beam. This is the traditional and oldest beam shape. It is still in production and used in building structural steel frames.
- W-beam or *Wide Flange* I-beam. This beam carries loads more efficiently than the S-beam on a weight-for-weight basis because it has a higher percentage of its total metal in its flanges where it can resist loading. They are also popular because more sizes of W-beams are available than S-beams, giving designers more choices. They are frequently used in building structural applications.
- H-beams. These are most often used for bearing piles and shoring. They are different from S- and W-beams because the flange and the web thicknesses are approximately equal, and the beam height and width are almost equal. They are also used to carry the exceptionally heavy loads held by the columns in high-rise buildings. They are particularly useful in resisting seismic loads because they offer more equal resistance to bending forces, both in line and perpendicular to the webs, than other I-beam shapes.

See Figure 6–83.

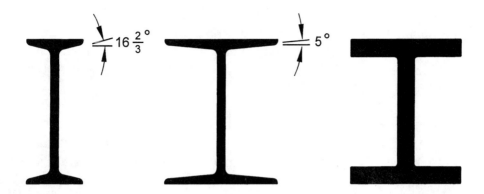

Figure 6–83. Common structural shapes: S-beam (left), W-beam (center), and H-beam (right).

Why are there more sizes of W-beams than I-beams?
Increasing the size and capacity of an S-beam is done by opening up the mill rolls. This makes both the flange wider and the web slightly thicker, Figure 6–84 (left). The additional material in the thicker web does not greatly increase the beam's strength despite its weight and cost. Because the manufacturing process for making W-beams permits *both* the web thickness and flange thickness to be varied independently, the designer can put more steel into the flange with or without adding to web thickness, Figure 6–84 (center and right).

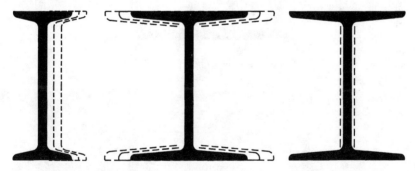

Figure 6–84. Increasing capacity of an S-beam is done by opening up the rolls (left), while a W-beam can have various flange sizes (center) and web sizes (right).

How are I-beams designated or *called out* on drawings?
In English units they are identified by their shape, nominal dimension (height) in inches, and weight in lb/ft. Similarly, in SI units they are identified by shape, nominal dimension in millimeters, and their weight in kN/m.

For example, in English units: **W36 × 848** calls for a wide flange I-beam with a nominal size or height of 36 inches and a weight of 848 lb/ft. In SI units, **W910 × 12.4** indicates a wide flange I-beam with a nominal height of 910 mm and a weight of 12.4 kN/m. To show that 47 of these pieces are needed, we write: **47-W36 × 848**.

What are the other common steel sections?
They are the Tees (or T-sections), angles, and channels. Large structural Tees are made by cutting an I-beam in half; this results in them having a flat edge on the bottom of the T-section. They are identified as either ST-sections or WT-sections depending on whether they were cut from an S-beam or a W-beam. Small T-sections, typically with legs of 4 inches or less, are initially rolled as Tees and have rounded bottoms. See Figure 6–85.

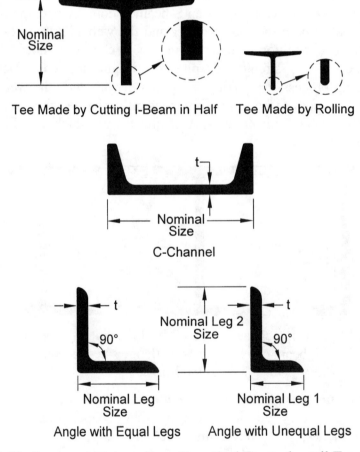

Figure 6–85. Common steel sections: Structural Tee and small Tee (top row), C-channel (middle row), angles (bottom row).

WELDING FABRICATION & REPAIR

How are these sections called out on drawings?
- Tees, by their nominal dimension and weight in lb/foot. For example:

  7-T 12 × 73 × 40'0"

  calls for seven lengths of T-section with a nominal size of 12 in, a weight of 73 lb/ft and a length of 40 ft.
- Angles, by their nominal leg dimensions, thickness, and length. For example:

  5-Ls 2 × 3 × 1/4 × 21'6"

  requires 5 pieces of L-section with 2-inch by 3-inch legs and ¼-inch thickness with an overall length of 21'6".
- Channels, by their nominal dimension and weight/length. For example:

  17-C 6 × 8.2 × 38'2"

  lists 17 lengths of C-channel with a 6-inch nominal size, a weight of 8.2 lb/ft and a cut length of 38'2".

How are square and rectangular tubular steel, pipe, and plate called out on drawings?
- Tubular steel, by the dimensions of one side, the dimensions of the other side, wall thickness and length. For example:

  13-T.S. 3" × 4" × 1/4 × 10'

  identifies 13 pieces of tubular steel 3-inch by 4-inches on its sides and with ¼-inch walls and a 10-foot cut length. See Figure 6–86. While *round* steel tube or steel tubing is available and widely used in mechanical applications, it is not often used in structural work.

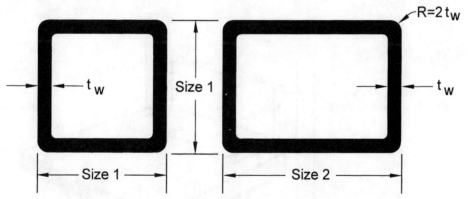

Figure 6–86. Tubular steel sections.

- Pipe is called out by:
  - Standard Pipe: (Nominal inside diameter) inches Std. Pipe × (length) feet. For example:

    6" Std. Pipe × 15'3"

- Extra Strong Pipe: (Nominal inside diameter) inches X-Strong Pipe × (length) feet. For example:

    12" X-Strong Pipe × 12'6"
- Double Extra Strong Pipe: (Nominal inside diameter) XX-Strong Pipe × (length) feet. For example:

    4" XX-Strong Pipe × 8'2"
- Plate sizing, by width, thickness, and length all in inches. For example:

    60" × 1/4" × 246"

On drawings plate steel is usually called for by writing the plate symbol, the thickness in inches, the width in inches, and the length in feet and inches. For example:

1/4" × 60" × 20'6"

Hooking Welds around Corners

When welding on structural steel, there are often questions about whether to hook fillet weld beads around a corner at the end of the bead. What is the general rule?

If bringing the weld around a corner maintains the size of the weld cross-section, then carry the bead around the corner, if not, don't.

The top seat angle connection, Figure 6–87, is designed to flex and give. Tests have shown that carrying the weld bead down the side of the angle for about ¼ of its length provides the greatest strength and movement before failure. Other examples of the application of this rule are shown in Figures 6–88 through 6–90.

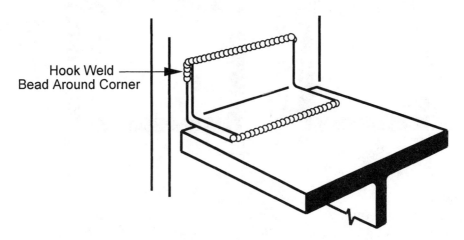

Figure 6–87. Hooking on top connection angle.

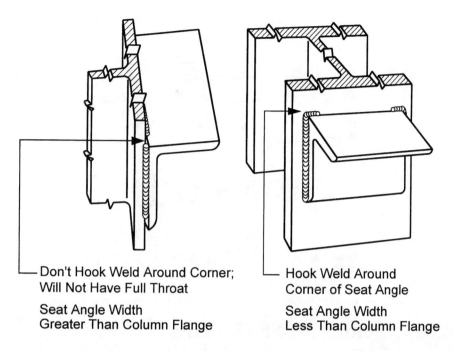

Figure 6–88. Hooking on seat angle fillet weld.

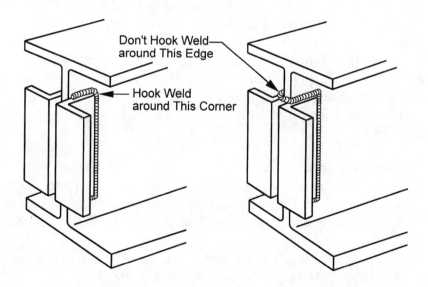

Figure 6–89. Hooking on beam to column connections.

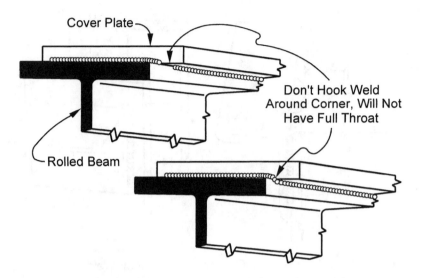

Figure 6–90. Hooking on beam cover plate.

Weld Cracking on Thick Plates

When placing fillet welds on thick plates, the welds sometimes crack. How can this cracking be avoided?

On thick plates with large welds, if there is metal-to-metal contact prior to welding, there is no possibility of plate movement. As the welds cool and contract, all of the shrinkage stresses must be taken up in the welds. In cases of severe restraint, this may cause the welds to crack, especially in the first pass on either side of the plate, Figure 6–91 (left). By allowing a small gap between the plates, the plates can 'move in' slightly as the weld shrinks. This reduces the transverse stresses in the weld. Heavy plates should always have a minimum of 1/32-inch (0.8 mm) gap between them, if possible 1/16-inch (1.6 mm) gap.

This gap can be obtained by means of:
1. Insertion of spacers made of soft steel wire between the plates, Figure 6–91 (center). The soft wire will flatten out as the weld shrinks. If copper wire is used, care should be taken that it does not mix with the weld metal.
2. A deliberately rough flame-cut edge. The small peaks of the cut edge keep the plates apart, yet can squash out as the weld shrinks.
3. Upsetting the edge of the plate with a heavy center punch. This acts in a similar manner to the rough, flame-cut edge.

The plates will usually be tight together after the weld has cooled, Figure 6–91 (right).

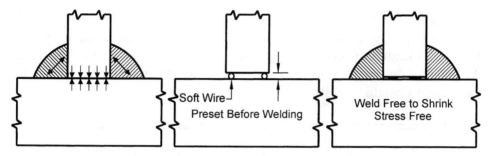

Figure 6–91. Using soft wire as a spacer to prevent weld cracking on thick plates.

Expanded Beams

What are expanded beams and why are they useful?

Expanded beams are made by cutting an I-beam through its web, adding a spacer to the web, and welding it all back together. Figure 6–92 shows a beam cut along the middle of its web, and the welding of an additional piece of steel into its web. This deepens the beam, increasing both its stiffness, and load-carrying ability.

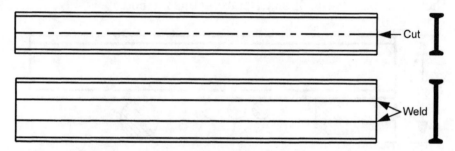

Figure 6–92. Expanded beam by adding a to its web.

If a beam is expanded by cutting the I-beam web on an angle, a tapered beam results, Figure 6–93. This beam has the most bending strength where it is most needed, in the center of the span. Such a design also lowers total steel load requiring support by the lower members of the frame.

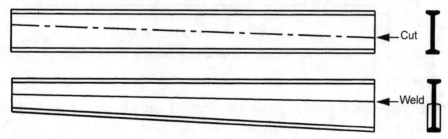

Figure 6–93. Expanded tapered beam.

If the web is cut in a zig-zag pattern along its center line and one of the two halves is turned end for end, an open-web expanded beam results, Figure 6–94. The resulting beam is deeper, stronger, and stiffer than the original. Starting the design with a lighter member realizes an immediate savings in material and handling costs.

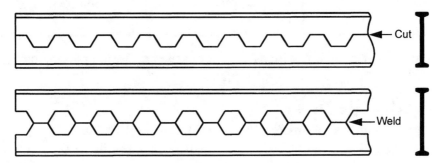

Figure 6–94. An expanded beam cut on zig-zag pattern.

Additional savings in steel and vertical space can be realized by using the openings in the beam for pipes, electrical conduit, and ductwork, Figure 6–95. There is also no waste of material using this method.

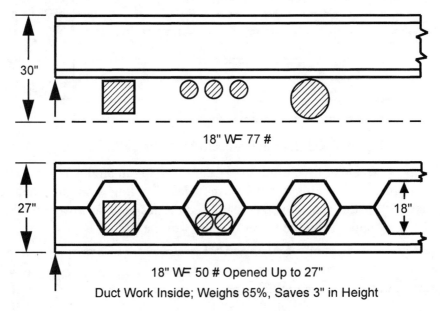

Figure 6–95. Steel and vertical height savings from open-web beams.

Another variant is a tapered open-web beam, Figure 6–96. All these designs are easily made using oxyacetylene cutting torch and a template.

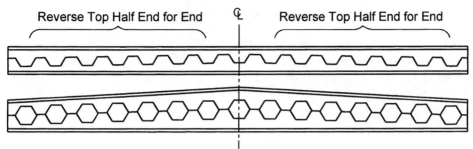

Figure 6–96. Tapered open-web design offers both steel and height savings.

Base Plate Designs

What are some typical base plate designs for structures?

Base plate designs are determined by the height of the structure, its floor loading, the underlying soil, and foundation strength, as well as wind and seismic forces, Figure 6–97.

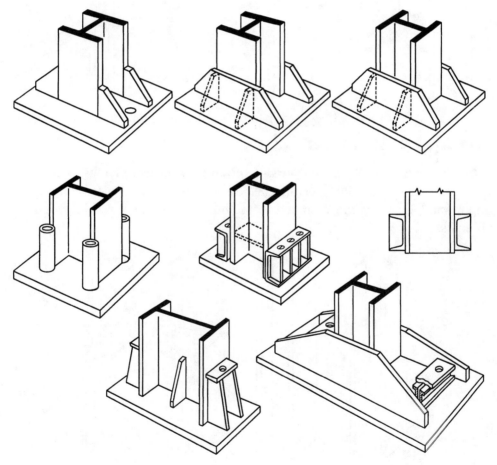

Figure 6–97. Typical base plate designs.

Beam-to-Column Connections
What are some typical beam-to-column connection methods?
Many factors determine the optimum beam-to-column connection design. The strongest connections require field welding, Figure 6–98 (left). But it is often possible to perform most of the welding in the shop and make the final field connections with high-strength structural steel nut and bolts, Figure 6–98 (center and right). This is convenient as it speeds field erection, and avoids weather-related welding problems.

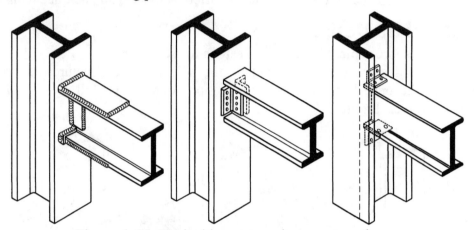

Figure 6–98. Typical beam-to-column connections.

Column Splice Designs
What are some common methods of splicing columns to each other?
There are many methods. Building floor, wind, and seismic loads are important factors. Some connection methods avoid field welding and use bolts instead. See Figure 6–99.

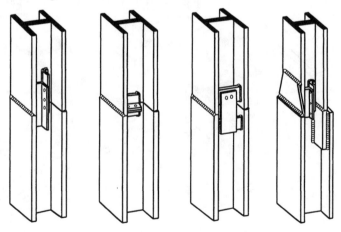

Figure 6–99. Typical column splices with welds. (continued on next page)

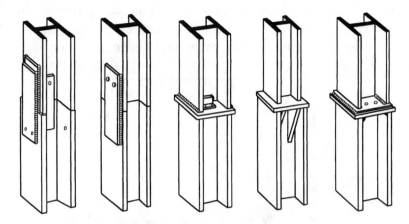

Figure 6–99. Typical column splices. (continued)

Column Supports and Guy Attachments

What are some effective ways to provide guy wire attachments for columns?
See Figure 6–100.

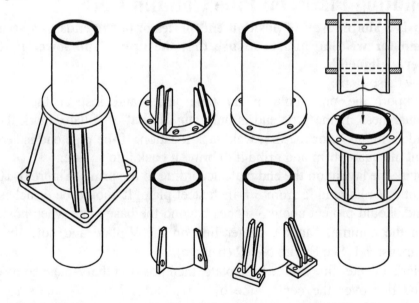

Figure 6–100. Typical designs for attaching and guying columns.

Weld Access

What are the functions of weld access holes?
There are two purposes. First, they provide the weldor access to the weld joint. Second, the weld access hole prevents the interaction of the residual stress field of the flange and the web. The interaction of the longitudinal

shrinkage of the web and flange weld, as well as the transverse shrinkage of the flange weld, creates triaxial stress that can crack the flange weld. The weld access hole eliminates this stress and reduces the chances of weld cracking, Figure 6–101.

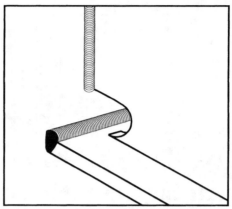

Figure 6–101. Weld access hole reduces chances of weld cracking by eliminating triaxial stresses on the flange weld.

Mounting Plates on Pipe Column Ends

What is a simple way to position end plates onto the ends of a structural column for welding and to assure that the upper and lower plates are properly aligned?

Follow these steps:
1. Support the pipe column on pipe jacks and inspect the ends for squareness. If the pipe end is only slightly out of square, mark the high side with an X for the first weld tack location. If the pipe end is seriously out of square, trim and grind it to bring it back into square.
2. Mark the layout on the end plate according to the prints, Figure 6–102 (a).
3. Cut a V-plate guide from 3/8-inch steel plate. It should be 6 inches wide and should project about 3 inches beyond the base plate when positioned on the column. Mark a center line on the V-plate and note the exact dimension t. See Figure 6–102 (b and c).
4. Mark a center line on the base plate, then position the V-plate so its center line lies over the center line of the plate and the V-plate's top edge projects the distance t above the base plate, Figure 6–102 (c). Clamp the V-plate to the base plate, Figure 6–102 (d).
5. Hang the base plate over the end of the column to position the plate and tack the plate in place, and then apply the final weld.
6. The next objective is to tack the unwelded base plate onto the other end of the pipe column so that the top and bottom plates are aligned, not rotated with respect to one another. To do this, put a torpedo level on the

WELDING FABRICATION & REPAIR

top edge of the previously welded base plate (left side of Figure 6–102 (e), and rotate the column, to bring this edge into level, Figure 6–102 (e).
7. Repeat step (4) above with the remaining base plate and hang it into welding position. Use a torpedo level to bring its top edge into level.
8. Tack the base plate and apply the final weld.

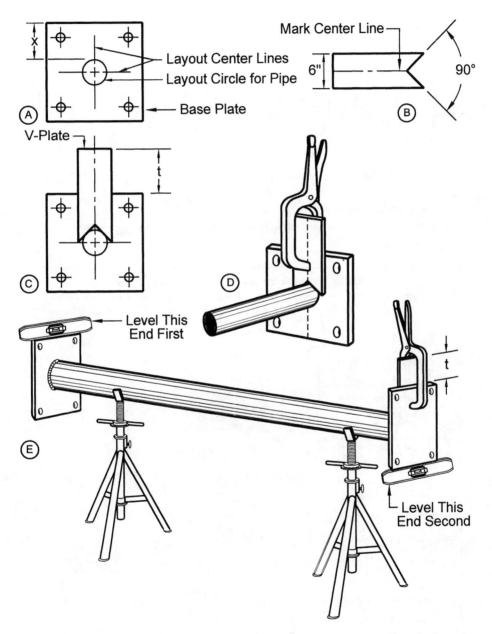

Figure 6–102. Steps in mounting base plates onto a structural pipe.

Hangers for Rolled Shapes

What are some ways to hang threaded rods from rolled structural shapes without welding on the shapes?
See Figure 6–103.

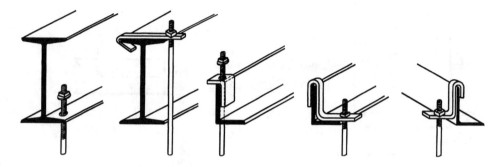

Figure 6–103. Methods of hanging threaded rods from structural shapes.

Welding Rebar

What are the standard ways of joining rebar by welding?
See Figure 6–104.

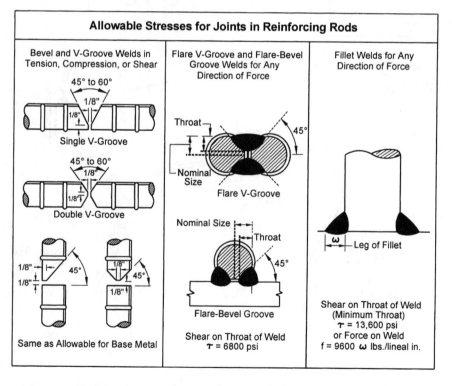

Figure 6–104. Methods of welding rebar.

Adding Plates to a Section

What are the common guidelines for adding plates to a section or sections?

See Figure 105.

Plate to a Rolled Shape, or
2 Plates in Contact with Each Other

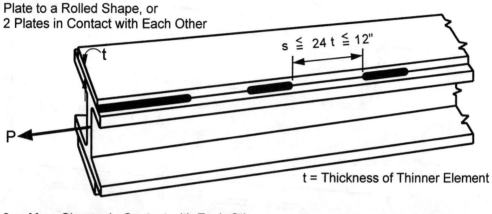

2 or More Shapes in Contact with Each Other

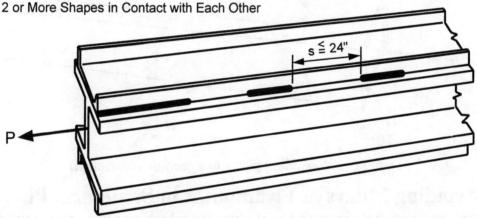

2 or More Shapes or Plates, Separated by Intermittent Fillers

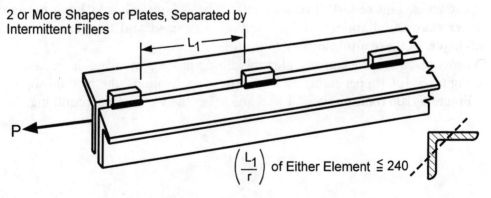

Figure 6–105. Adding plates to a section. (continued on next page)

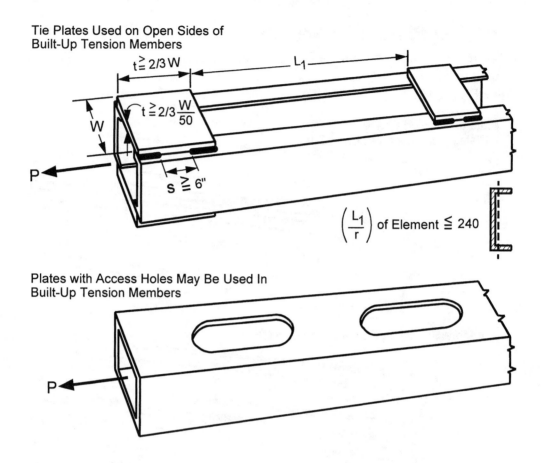

Figure 6–105. Adding plates to a section. (continued)

Avoiding Miters & Fishmouths in Structural Pipe
Making joints between structural pipes (or tubes) requires careful fitting of the ends, Figure 6–108 (upper left). These joints should have no gaps larger than the diameter of the welding wire used and they can be labor intensive. Is there another way to do this?

In many instances, we can side step this fitting process by using the method of Figure 6–106 (upper right). In lower stress applications, the method shown in Figure 6–106 (bottom) works well and saves the work of fishmouthing.

WELDING FABRICATION & REPAIR

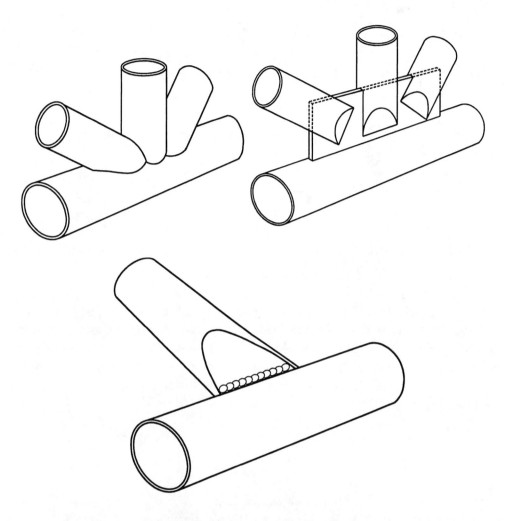

Figure 6–106. Three methods of joining structural pipes or tubes.

Section VIII – Oxyfuel & Lance Cutting

Gas Pressures & Cutting Tip Sizes

What are the usual regulator pressures and tip sizes for common steel thicknesses?

See Table 6–2. Note that the *Number 0* cutting tip will be used most often.

Steel Thickness		Pressure (psi)		Tip No.	Drill No. diam. (inches) of O_2 Orifice
inches	mm	Oxygen	Acetylene		
1/8 – 1/4	3.2-6.4	25	5	00	67 (0.032)
3/8 – 5/8	10-16	35	8	0	60 (0.040)
3/4 - 1	19-25	45	10	1	56 (0.047)
1 - 1½	25-38	50	12	2	53 (0.060)
2 - 2½	51-64	55	12	3	50 (0.070)
3 - 4	76-102	60	15	4 or 5	45 (0.082) or 39 (0.100)

Table 6–2. Oxyfuel pressure settings and tip sizes vs. carbon steel thickness.

What is the number one cause of cutting problems?

A dirty cutting tip. Remove the tip from the torch and use tip cleaners to make sure all the orifices are unobstructed and clean of carbon. Avoid enlarging or "belling" the orifices; this will destroy the tip. Next, holding the tip perpendicular to fine emery cloth on a flat, smooth surface, move the tip back and forth several times. This will remove any burrs from the orifices and remove any carbon deposits from the tip's surface. Look for a shiny, flat surface, Figure 6–107.

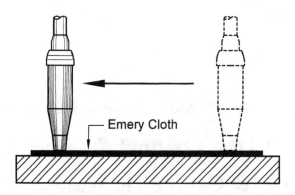

Figure 6–107. Cleaning the end of a cutting tip on emery cloth.

Diagnosing Oxyfuel Cutting Problems

You have cleaned your cutting torch tip and still have cutting problems. What should you do?

Match the cut with the drawings in Figure 6–108 to identify the problem source.

WELDING FABRICATION & REPAIR

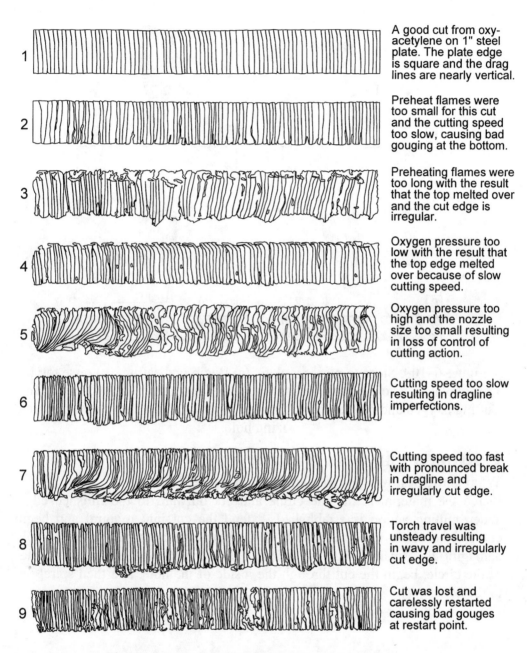

1. A good cut from oxy-acetylene on 1" steel plate. The plate edge is square and the drag lines are nearly vertical.

2. Preheat flames were too small for this cut and the cutting speed too slow, causing bad gouging at the bottom.

3. Preheating flames were too long with the result that the top melted over and the cut edge is irregular.

4. Oxygen pressure too low with the result that the top edge melted over because of slow cutting speed.

5. Oxygen pressure too high and the nozzle size too small resulting in loss of control of cutting action.

6. Cutting speed too slow resulting in dragline imperfections.

7. Cutting speed too fast with pronounced break in dragline and irregularly cut edge.

8. Torch travel was unsteady resulting in wavy and irregularly cut edge.

9. Cut was lost and carelessly restarted causing bad gouges at restart point.

Figure 6–108. Cutting problems and their causes.

Piercing a Plate

What procedure is used to pierce a plate?
See Figure 6–109.

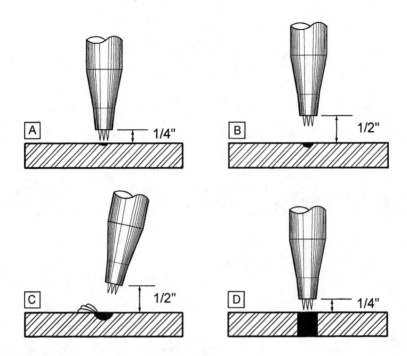

Figure 6–109. Steps to pierce a plate: (a) Begin preheating, (b) raise torch from plate to avoid cutting splatter onto tip, (c) tilt torch off vertical, open cutting oxygen and pierce plate, (d) move torch back to vertical and complete sizing hole.

Cutting a Circle in a Plate
What is the best way to cut a circle from a plate?
Pierce the material *inside* the circle and cut outward to the finished edge. When cutting action is established, extend the cut into a spiral and begin cutting the circle itself, Figure 6–110. When cutting small circles, avoid damaging the finished edge by drilling a 1/4-inch (6 mm) hole in the center of the circle, begin the cut through the inside of the hole, and then spiral out to the edge.

WELDING FABRICATION & REPAIR

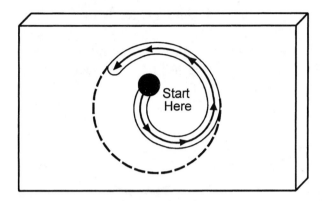

Figure 6–110. Cutting a circle in plate.

Cutting I-Beams

What is the easiest way to cut an I-beam to length?
Make your cuts in the sequence shown in, Figure 6–111. By making the web cut vertically up, the falling slag does not clog the tip as it would in a vertically down cut. Also, by making the final cuts along the top of the flange, the beam end is securely supported until it falls away from the main section.

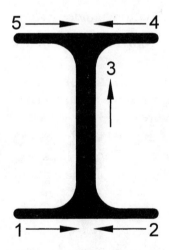

Figure 6–111. Severing an I-beam.

Aids for Straight Cuts on I-Beams

How can I-beams be accurately and smoothly cut with a hand-held torch?
Make cutting guides for your torch, Figure 6–112. Use a T-square style guide for the top and bottom of the flanges, and a cut-to-fit guide to cut the web.

Note that this guide has rounded edges and fits tightly against the web.

Some guides are made thick enough to let the nut of the welding torch ride *along* the guide and eliminate the need for the weldor to control the torch height from the beam surface.

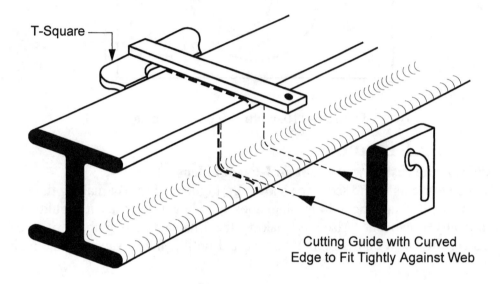

Figure 6–112. Shop-made cutting guides insure straight cuts on I-beams.

How can a frozen nut be removed from a bolt without damaging the bolt threads?
- Place the nut in a horizontal position, and using the minimum cutting torch size and oxygen pressure, begin to heat the nut on a corner of its hexagon points. See Figure 6–113 (a).
- When the point metal turns red, apply the oxygen and begin cutting. At the same time, turn the cutting torch head so it is parallel to the threads. We do not want to let the cutting flame cut down into the bolt threads. See Figure 6–113 (b).
- Continue to cut a slot down through the bolt until just above the bolt threads.
- Finally, quickly heat and flush away the remaining nut metal. The bolt threads will not be burned if they were not allowed to become red hot. See Figure 6–113 (c).
- Rotate nut/bolt combination 180 degrees and repeat the above steps. See Figure 6–113 (d & e).

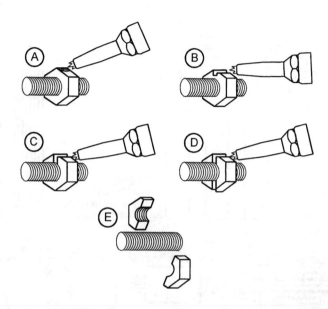

Figure 6–113. Burning off a nut.

Oxygen Lances & Burning Bars

What is an oxygen lance, what is it used for, and how does it work?
A typical oxygen lance system consists of:
- Oxygen source, either manifolded oxygen cylinders, or liquid oxygen cylinder and vaporizer. These are required to supply the high-volume of oxygen required; a single oxygen cylinder is not adequate.
- High-volume oxygen regulator capable of 40 ft^3/min (68,000 l/hr) at 150 psi (10 bar) minimum.
- Oxygen lance hose with a 3/8-inch diameter minimum; 1/2-inch diameter is needed for hose lengths over 100 feet (30 m).
- Lance pipe, which is typically schedule 40 black steel pipe, from 3/16- to 1-inch (4.7 to 25.4 mm) OD depending on the application. The pipe may be ERW or CW and the lance pipe length is usually about 10.5 feet (3 m), but may run from 3 to 21 feet (1 to 7 m).
- Valve and pipe holder assembly usually consisting of a ball valve and hardware to connect the valve to the lance pipe. There may also be a heat shield and flash back arrestor.

See Figure 6–114.

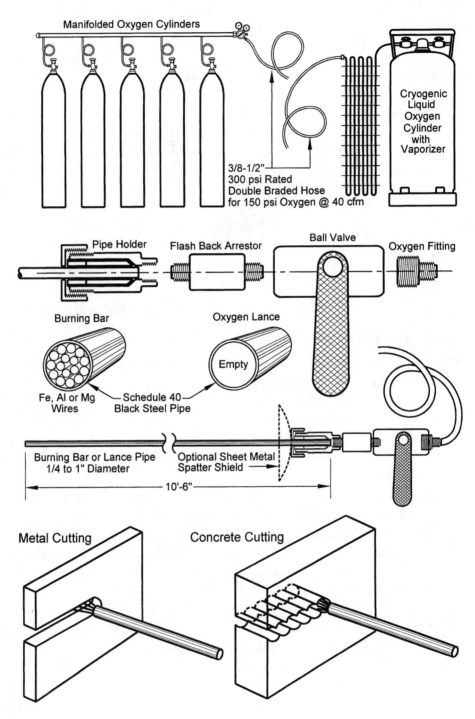

Figure 6–114. Oxygen lance and burning bar.

Oxygen lances are most often used for cutting cast iron and steel in foundries and steel mills where the materials to be cut are *already* red hot from processing and need no additional preheating to begin lancing. There are many other production and maintenance applications in foundries and steel mills. For example, if equipment failure allows metal in production to cool in a ladle, an oxygen lance can cut up the solidified metal so it can then be removed in pieces.

When material to be cut is *not* already red hot, a second weldor usually preheats the start of the cut with an oxyacetylene torch. When the starting point is red hot, the lance operator positions the lance over the preheated spot and opens the oxygen valve. As the steel of the lance pipe burns and is consumed in the oxygen stream, it provides heat to melt and ignite the work. As the lance and work metals burn in oxygen, they form fumes and slag that is blown from the kerf producing a cut.

What is the difference between an oxygen lance and a burning bar?
An oxygen lance uses a length of *empty* steel pipe to contain and direct the oxygen stream and to provide heat from the oxidation of the pipe steel to the work. A burning bar, more correctly called an *exothermic burning bar*, adds iron, aluminum, or magnesium wires or rods *inside* the steel lance pipe. These rods provide substantially more heat to the work than an oxygen lance and cut much more rapidly. They will cut virtually any metal, concrete, or refractory (fire-brick). Temperatures between 6600 and 8000°F (3650 and 4430°C) have been recorded on burning bars and 12,000°F (6650°C) in high performance burning bars. See Table 6–3 for melting points of materials commonly cut with oxygen lances and burning bars. The same oxygen supply equipment works for both the lance and burning bar, but burning bars require oxygen pressures as high as 250 psi (17 bar), so a special double-braided oxygen hose with a rated working pressure of 300 psi (21 bar) is required.

Burning Bar/Material	Output Temperature or Melting Point	
	°F	°C
Exothermic Burning Bar	6600-8000	3650-4430
Exothermic High Temperature Burning Bar	12000	6650
Steel	2850	1565
Copper Alloys	1985	1084
Nickel Alloys	2650	1455
Concrete	5200	2870

Table 6–3. Melting point comparison for burning bars versus materials.

In general oxygen lances work better on iron and steel already red hot from processing, while burning bars work better cutting cold metals and concrete.

Do oxygen lances and burning bars cut in a continuous line?
Both cut metals in a continuous line, but sever concrete, masonry, and refractory material by melting a series of overlapping holes much like chain drilling in metal. See Figure 6–114 (bottom).

How fast is a typical 10.5-foot (3 meter) burning bar consumed, how much oxygen will it use, and how much concrete can a single 10.5-foot bar cut?
It is consumed in 4 to 4½ minutes and uses 100 ft^3 (2832 l) of oxygen. In normal operation it will use 1200 ft^3/hour (34,000 l/hr). A 3/8-inch diameter burning bar will melt through a foot of concrete per minute, or 24 to 30 inches (0.6 to 0.76 m) of 2-inch (5 cm) diameter holes.

How much steel will the same size burning bar cut?
It will cut about 17 feet of 2-inch thick carbon steel, or 14 feet of 3-inch thick stainless steel.

Do reinforcing rods in the concrete diminish the cutting ability of the burning bar?
No, they actually speed up the cutting action because they provide iron to burn in the oxygen stream.

What factors affect the ability in burning bars to cut concrete?
Concrete with smaller aggregate is easier to cut than concrete with large aggregate.

How does the action of lances and burning bars differ when cutting copper and concrete?
Lances and burning bars actually oxidize iron and steel and blow away the oxides in the form of slag. Other materials like copper and concrete are cut by melting them and blowing the melted liquid material from the cutting zone.

What are the advantages and disadvantages of the oxygen lance and the exothermic burning bar?
The advantages are very high-speed cuts with no mechanical vibration, both advantages in demolition and rescue work. In some cases, there is no other effective way to cut large masses of material.

Welding Fabrication & Repair

The disadvantages are that the lance and burning bar consume large volumes of oxygen and generate a lot of heat, light, sparks, and fumes. In heavy industry where oxygen is purchased in bulk and is relatively inexpensive, oxygen consumption is not an issue.

Do burning bars make faster cuts than oxyfuel torches?
In general, yes. For example, a 44 inch (1.1 m) diameter forged carbon steel generator shaft had to be cut into 40 inch (1.0 m) lengths to facilitate its removal during demolition. An oxyacetylene cutting torch took 14 hours to make each cut, while an exothermic burning bar took 45 minutes per cut. Each cut required nine burning bars 0.922 inch (23 mm) diameter and 10.5 feet (3.2 m) long using 80 ft^3/minute (2300 l/min or 136,000 l/hr). In many applications burning bars are the best solution from both a time and total cost standpoint.

Are there other important variations of burning bars?
Yes, there are designs to cut metals under water and others to perform welding under water too.

What personal safety equipment is needed for lance and burning bar operation?
Each operator must have:
- Safety glasses
- Hard hat with full face shield (#4 or #5 tint)
- Flame resistant suit including jacket, pants, and boot covers
- Fire resistant gloves

What are the major safety issues in operating lances and burning bars?
- Wear proper protective clothing to protect skin and eyes from heat, UV radiation, and sparks.
- Maintain proper oxygen pressure and volume.
- Stop using the system if the regulator or hose begins to freeze up.
- Never use the equipment alone.
- Have a fire watch or safety person standing by.
- Maintain adequate fresh breathing air to the operator and protect the operator from exposure to cutting fumes.
- Inspect all oxygen handling equipment to make sure it is free from grease and dirt. All equipment should be supplied by the manufacturer, cleaned for "oxygen service." Do not use pipe dope on oxygen equipment threads; use PTFE tape.

- Store all oxygen handling equipment in a clean area and away from oil and grease.
- Plug or tape shut oxygen equipment fittings when not in use to keep out dirt and insects.
- Be aware that while supplying high volumes of oxygen, many oxygen regulators may freeze up, halting cutting action.

Section IX – Industrial Fasteners

What are the five general categories of threaded fasteners?
See Table 6–4.

Type Fastener	Organization Issuing Specifications	Typical Applications
Commercial	Often none, but may be mismarked as SAE Grade	Household, wood, non-critical, use with caution
Automotive	Society of Automotive Engineers (SAE)	Automotive, off-road vehicles
Structural Steel	American Society for Testing Materials (ASTM)	Buildings & structures
Aircraft	US Military	Aircraft, missiles, high-performance vehicles like racing cars & boats
Metric	American Society of Mechanical Engineers (ASME), International Standards Organization (ISO), SAE, ASTM	US Automotive, widely used in Europe for most applications

Table 6–4. Fastener types, specifications, and uses.

Commercial Fasteners

What are commercial fasteners?
They are the fasteners—nuts, bolts, and washers—sold in local, hardware stores. They are the cheapest, lowest class of fastener. They are often made overseas. There is no way to determine which of the following conditions exist:
- Not made to meet *any* specification.
- May have SAE Grade markings, but not meet marked Grade specification.
- Have SAE Grade markings and do meet SAE Grade specifications.

As a result, fasteners purchased from local hardware stores should be used with great caution and only for non-critical applications. Their failure is not predictable.

SAE Fasteners
How do SAE specifications classify threaded fasteners?
They are divided into *SAE Grades* that run from Grade 1 to Grade 8. Higher Grade numbers indicate a stronger fastener. Table 6–5 shows the bolt head identification markings on SAE graded bolts with size range and strength information. If these fasteners are purchased from a reliable supplier with paperwork to support their manufacturer and SAE Grade, they are likely to meet design requirements.

Head Marking	Spec.	Materials	Bolt and Screw Size (Inches)	Proof Load PSI x 1000	Tensile Strength Minimum PSI x 1000
(plain hex head)	SAE Grade 1	Low or Medium Carbon Steel	1/4 - 1-1/2	33	60
(plain hex head)	SAE Grade 2	Low or Medium Carbon Steel	1/4 - 3/4 7/8 - 1-1/2	55 33	74 60
(3 radial lines)	SAE Grade 3	Medium Carbon Steel, Cold Worked	1/4 - 1-1/2 1/2 - 5/8	85 80	110 100
(3 radial lines)	SAE Grade 5	Low or Medium Carbon Steel, Quenched and Tempered	1/4 - 1 1-1/8 - 1-1/2 1-1/2 - 3	85 74 55	120 105 90
(3 radial lines)	SAE Grade 5.1	Low or Medium Carbon Steel, Quenched and Tempered with Lock Washer	Up to 3/8	85	120
(3 radial lines)	SAE Grade 5.2	Low Carbon Martensitic Steel, Quenched and Tempered	1/4 - 1	85	120
(5 radial lines)	SAE Grade 7	Medium Carbon Alloy Steel, Quenched and Tempered	1/4 - 1-1/2	105	133
(6 radial lines)	SAE Grade 8	Medium Carbon Alloy Steel, Quenched and Tempered	1/4 - 1-1/2	120	150

Table 6–5. SAE bolt head markings.

Structural Steel Fasteners

What are the most common structural steel fastener specifications?
They are the specifications for heavy hexagon bolts, nuts, and washers. They are usually offered in steel, weathering steel, and galvanized steel. Their diameters range from 1/2 to 1½ inches and are made to the ASTM specifications in Table 6–6.

Component	Specification	Description
Bolts	ASTM A325	Conventional, heavy-hex high-strength bolts in steel-to-steel construction
	ASTM A490	Slightly higher strength than ASTM A325 bolts.
	ASTM F1852	Twist-off type tension-control bolt (same strength as ASTM A325 and marked similarly)
Nuts	ASTM563	Heavy-hex nuts
Washers	ASTM F436	Hardened steel washers
Direct tension indicator (DTI)	ASTM F490	Direct tension compressible washer

Table 6–6. Structural steel fasteners and their specifications.

In what variations are structural steel fasteners offered?
The two principal choices are ASTM A325 steel with a tensile strength of 120,000 psi and ASTM A490 alloy steel with a tensile strength of 150,000 psi. Again there is a choice among steel, weathering steel, and galvanized steel.

There are also nuts for both ASTM A325 and ASTM A490. Because it is important to use nuts and bolts of matching specifications, structural steel fasteners have raised markings showing their specification and manufacturer, Table 6–7 and 6–8. The small n on the bolt head and nut markings indicates Nucor Steel's Fastener Division produced them.

CHAPTER 6 — WELDING PROBLEMS, SOLUTIONS & PRACTICES

Head Marking	Spec.	Materials	Bolt Size	Tensile Strength Minimum PSI x 1000
A325	ASTM A325 Type I	Medium Carbon Steel Quenched and Tempered	1/2" - 1" 1-1/8" - 1-1/2"	120 105
A325	ASTM A325 Type III	Weathering Steel Quenched and Tempered	1/2" - 1" 1-1/8" - 1-1/2"	120 105
A490	ASTM A490 Type I	Alloy Steel Quenched and Tempered	1/4" - 1-1/2"	150
A490	ASTM A490 Type III	Weathering Steel Quenched and Tempered	1/4" - 1-1/2"	150
A325M 8.8S	Property Class 8.8S ASTM A325M Type I ISO 7412	Metric Version of ASTM A325 Type I	M20, M22, M24	120

Table 6–7. Structural steel bolt markings, materials, sizes and specifications.

Nut Face Marking	Spec.	Materials	Bolt Size (Inches)	Proof Load PSI x 1000
(hex nut, C, n)	ASTM Type I Grade C	Alloy Steel	1/2 - 1 < 1	120 105
(hex nut, C₃, n)	ASTM A563 Type III Grade C3	Weathering Alloy Steel with Cu, Ni, Cr	1/2 - 1 < 1	120 105
(hex nut, DH, n)	ASTM Type I Grade DH	Alloy Steel	1/2 - 1	175
(hex nut, DH3, n)	ASTM A563 Type III Grade DH3	Weathering Alloy Steel with Cu, Ni, Cr	1/2 - 1	175

Table 6–8. Structural steel nut markings, materials, sizes and specifications.

Aircraft Fasteners

How do aircraft-type fasteners differ from commercial and SAE-type fasteners?

Here are some important differences. Aircraft fasteners are:
- Made to US Government/US Military specifications, not commercial or automotive specifications.
- They are much more expensive.
- Materials include corrosion-resistant, high-fatigue-resistant steel alloys, aluminum, and titanium.
- Bolts are heat-treated to 125,000 psi or better.
- Many come drilled for safety wires to prevent loosening under vibration.
- Available in a very wide variety of sizes and designs, including close-tolerance hex-head bolts, internal wrenching bolts, external wrenching

bolts, close-tolerance shear bolts, clevis bolts for shear applications, and eyebolts.
- Also available are a wide assortment of washers, nuts, and vibration-resistant nuts.
- Bolts have ID markings on heads.
- Specifications begin with the pre-fixes: *AN*, *NAS*, and *MS*.

Metric Fasteners

How are metric bolts typically called out?

An example is **M20 × 60, property class, 10.9**. Here the **M** indicates a metric fastener, the **20** represents the bolt's nominal diameter in mm, the **60** represents bolt length in mm, and the property class of **10.9** indicates the tensile and yield strength of the bolt metal.

What do metric bolt head markings indicate?

The raised bolt head markings show the manufacturer and the property class, which indicates the tensile and yield strength, Table 6–9.

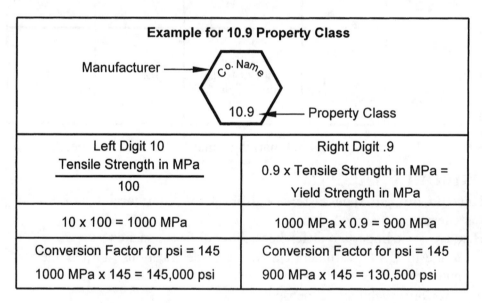

Table 6–9. Metric bolt head markings and their meaning.

In what variations are metric threaded fasteners offered, particularly for structural steel fasteners?

Table 6–10 indicates property class, head style, size ranges available, and their specifications. Table 6–11 shows property class, nominal sizes, materials, yield strength, tensile strength, and bolt head markings.

FASTENER DATA				
Basic Product	Product Type and Head Style	Available Size Range	For Thread and Dimension Details Refer To:	For Mechanical Property Details Refer To:
Metric Bolts	hex	M5-M100	ANSI/ASME B18.2.3.5M	ASTM F#568 ASTM F#486M ASTM F#738
	heavy hex	M12-M36	ANSI/ASME B18.2.3.6M	
	heavy hex structural	M12-M36	ANSI/ASME B18.2.3.7M	ASTM A#325M ASTM A#490M
	hex transmission tower	M16-M24	IFI 541	IFI 541
Metric Screws	hex cap	M5-M100	ANSI/ASME B18.2.3.1M	ASTM F#568 ASTM F#468M ASTM F#738
	heavy hex	M12-M36	ANSI/ASME B18.2.3.3M	
	hex flange	M5-M16	ANSI/ASME B18.2.3.4M	
	heavy hex flange	M10-M20	ANSI/ASME B18.2.3.9M	
Metric Nuts	hex, style 1	M1.6-M36	ANSI/ASME B18.2.4.1M	ASTM A#563M ASTM F#467M ASTM F#836M ASTM A#194M
	hex, style 2	M3-M36	ANSI/ASME B18.2.4.2M	
	slotted hex	M5-M36	ANSI/ASME B18.2.4.3M	
	heavy hex	M12-M100	ANSI/ASME B18.2.4.6M	

Table 6–10. Metric fasteners, product type, available size range, and specifications.

Property Class Designation	Nominal Size of Product	Material and Treatment	Mech. Requirements		Property Class Ident. Marking
			Yield Strength MPa, Min.	Tensile Strength MPa, Min.	
4.6	M5-M100	low or medium carbon steel	240	400	4.6
4.8	M1.6-M16	low or medium carbon steel, fully or partially annealed	340	420	4.8
5.8	M5-M24	low or medium carbon steel, cold worked	420	520	5.8
8.8	M16-M72	medium carbon steel, quenched and tempered	660	830	8.8
A325M Type 1	M16-M36				A325M 8S
8.8	M16-M36	low carbon boron steel, quenched and tempered	660	830	8.8
A325M Type 2					A325M 8S
A325M Type 3	M16-M36	atmospheric corrosion resistant steel, quenched and tempered	660	830	A325M 8S3
9.8	M1.6-M16	medium carbon steel, quenched and tempered	720	900	9.8
9.8	M1.6-M16	low carbon boron steel, quenched and tempered	720	900	9.8
10.9	M5-M20	medium carbon boron steel, quenched and tempered	940	1040	10.9
10.9	M5-M100	medium carbon alloy steel, quenched and tempered	940	1040	10.9
A490M Type 1	M12-M36				A490M 10S
10.9	M5-M36	low carbon boron steel, quenched and tempered	940	1040	10.9
A490M Type 2	M12-M36				A490M 10S
A490M Type 3	M12-M36	atmospheric corrosion resistant steel, quenched and tempered	940	1040	A490M 10S3
12.9	M1.6-M1000	alloy steel, quenched and tempered	1100	1220	12.9

Table 6–11. Mechanical requirements for metric carbon steel fasteners.

Bolting Basics
What are the parts of a bolt?
See Figure 6–115.

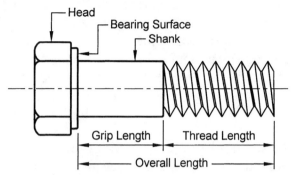

Figure 6–115. Parts of a bolt.

What are the general guidelines for using structural nuts and bolts?
- Fasteners must be kept clean and dry.
- Dirty fasteners must be cleaned and relubricated.
- Production lot number of bolts, nuts, and washers should be recorded and the fasteners of the same kind separated by lot. The mixing of different lots is not permitted; only pretested combinations may be used.
- Galvanized nuts and bolts must be kept together as an assembly as supplied by the manufacturer to assure proper thread fit after galvanization.
- Joints must be snugged and tightened in a systematic manner.
- Snugging is to begin at the most rigid part of the joint and proceed to the most flexible end.
- Pretensioning follows snugging, if the joint requires it.
- All plain or black bolts must be lubricated. The most effective way to do this is to put lubricant on the threads of the nut and bolt and the inside face of the nut which will bear on the washer.
- A special lubricant is required for galvanized fasteners.

Exactly what does snugging mean?
Snugging, the snug tight condition, is defined as the "tightness attained by either a few hits of an impact wrench or the full effort of a worker with an ordinary spud wrench that bring the connected plies into firm contact." This is the most recent AISC LRFD Specification (12/1/93), §J3.1.

What important application rule must be followed when using nuts and bolts to secure two or more layers of material together, particularly in structural applications?

Bolts are designed to work in tension; they are not designed to resist shear forces. To put and keep the bolt in tension under all anticipated loads, the bolt must be properly *pretensioned* (or pulled up) with its nut and washer when installed. A properly tensioned nut and bolt pair pulls the materials between them together under enough force that the friction developed between the plies clamps them from moving with respect to each other. The bolt is *never* put in shear.

In general, bolts should not be used as bearings or locator pins. There are a few aircraft-type bolts that are specifically designed for use in shear, but they are not used in structural applications.

How is pretensioning accomplished?

There are four ways to do this:

- *Turn of the bolt method* depends on having the steelworker tighten the bolt so many degrees after it is "snugged up" and the joint is closed.
- *Torque on the bolt method* uses a torque wrench to pull up the nut to a predetermined torque.
- *Direct tension indicators (DTIs)*—These special washers show how much tension is on the bolt, Figure 6–116. The tension on the bolt is transferred to the washer's dimples. A feeler gauge determines how much the dimples have been flattened, which directly indicates bolt tension, Figure 6–117. Typically the washers start with a 0.030-inch gap and reach minimum pretension at either a 0.015- or 0.005-inch gap.

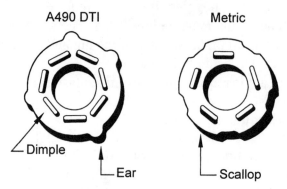

Figure 6–116. Direct tension indicators are available in both English and Metric sizes.

WELDING FABRICATION & REPAIR

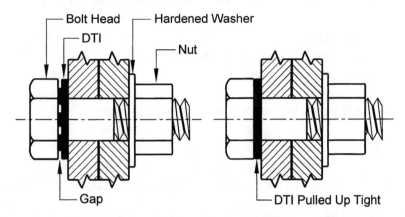

Figure 6–117. How DTIs work: Note the gaps between base of DTI and bolt head before tightening (left) and the reduction in gaps to either 0.005 or 0.015 inches, indicating proper bolt tension (right).

- *Snap-off bolts* have an external driving spline and are designed to fracture at their driving point at a predetermined torque, Figure 6–118. These bolts are pulled up with an electrically- or air-driven tool.

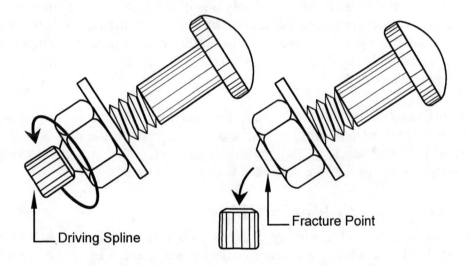

Figure 6–118. Snap-off bolts have an external driving spline segment designed to snap off at a predetermined torque.

What is the objective of these four bolt pretensioning methods?
The sole objective of *all* methods of pretensioning structural bolts is to put a minimum pretension on the bolt, typically about 40,000 psi.

What are the relative advantages of each of these four pretensioning methods?
Both the turn-of-the-nut and torque-on-the-bolt methods depend on a the results of tests with a Skidmore-Wilhelm Bolt Tension Indicator, commonly called a "Skidmore." These tests are conducted on the work site each day to determine the rotation or the torque required to develop a bolt pretension level for a particular fastener lot, as measured by the Skidmore.

The turn-of-the-bolt method depends entirely on the results of daily on-site tests to establish the relationship between nut rotation and bolt tension. It also depends on the care of the steelworker in applying the required nut rotation.

The fundamental weakness of the torque-on-the-bolt method is that roughly 90% of the torque applied to pretension the bolt goes to overcome nut-to-washer and thread-to-thread friction. Only 10% goes to pretensioning. Although the Skidmore has accuracy in the 1% range, this approach depends entirely on the assumption that the coefficient of friction between the nut and washer and the nut threads and bolt threads will be the same for all fasteners—a giant leap of faith.

The Direct Tension Indicator is the only one of the four methods mentioned above that actually measures bolt tension on every bolt installed. It is reliable and accurate. Some DTIs have a high-visibility orange silicone compound inside their dimples that squirts out when the bolt reaches proper tension. No feeler gauges are used.

Snap-off bolts suffer from the weakness that they depend on measuring applied torque, not bolt tension. Variations in friction on the nut, bolt, and washer faces produce large swings in bolt pretension. These bolts are very convenient to use as they are tensioned with a power tool.

Brazing and Soldering Tips
You have a Bernz-O-matic®-type torch with a hose and find it handy, except that there is no safe and convenient place to put a lit or hot torch if you must set it down. What should you do?
From scrap steel make the cylinder and torch holder shown in Figure 6–119. This holder will hold a lit torch without starting a fire. It will also let you adjust the regulator knob easily by preventing the cylinder from turning and falling over.

WELDING FABRICATION & REPAIR 247

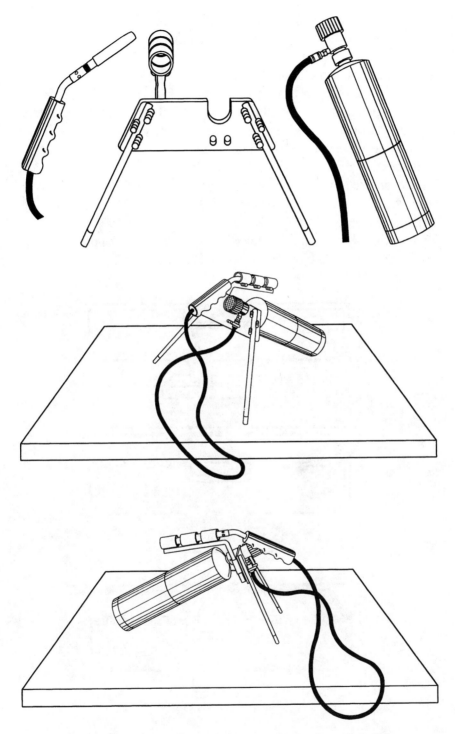

Figure 6–119. Shop-made cylinder and torch holder.

You have a 1/4-inch (6.3 mm) twist drill bit and want to braze it onto a piece of 1/4-inch drill rod to make an extension bit. How should you do this?

Use an oxyfuel torch to make the braze. Here are the steps:
1. Make up the brazing fixture as shown in Figure 6–120 (a, b & c).
2. Square the ends of the drill and drill rod extension on brazing faces.
3. Clean the drill bit and drill rod with emery cloth down to clean, bare metal and wipe off with acetone or alcohol.
4. Secure the drill bit and drill rod as shown in Figure 6–121 (a). Additional clamping may be needed to insure the parts do not move. The joint must be no wider than 0.005 inches (0.12 mm) or the joint will not have maximum strength.
5. Apply flux, heat, and add silver-based braze metal to the parts.
6. When cool, smooth the joint with a file or grinder.

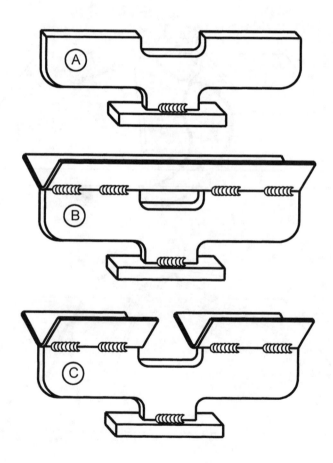

Figure 6–120. Making a brazing fixture: (a) Making the frame, (b) adding the angle iron bed, (c) cutting the opening in the angle iron bed to permit brazing.

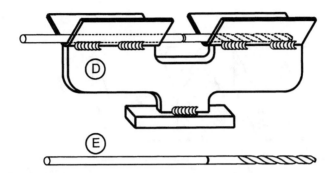

Figure 6–121. Brazing an extension on a twist drill: (a) drill rod and drill ready to braze, (b) drill with brazed extension.

Brazing Castings

A large casting has broken and you must braze welding to repair it. How is this done?

Here are the steps, Figure 6–122:
1. Remove all oil, grease, paint, and dirt from the crack faces and back 2 inches (5 cm) from each face. See the Cleaning section in Chapter 1 for methods.
2. Grind a V-groove on the top and bottom joint edges, but do not go all the way through to the middle. Leave some of the broken face in place to provide precise joint alignment. If the repair is to a *partially* cracked casting, drill stress relief holes 3/8-inch (10 mm) diameter at the end of the crack(s). Then grind a V-groove about 1/3 of the way to the center of the casting from each side.
3. Preheat the entire casting to 500-900°F (260 to 485°C). One or more rosebud tips will be needed to do this. This heating will reduce the chances of warping and cracking.
4. Apply borax-boric acid flux to the joint area. A flux with brass spelter (zinc particles) will facilitate tinning the joint faces.
5. Begin at one end and heat the joint to a dull red; then add bronze filler rod. Make sure the joint faces are thoroughly tinned and work your way along the joint to the other end. Several passes may be needed for thick castings.
6. Alternate sides when adding filler metal.
7. Complete the repair by reinforcing or crowning the joint so it is about 1/2-inch above the face of the casting.
8. Either bury the casting in sand to retard its cooling or cool it in an oven. Not doing so risks cracking.
9. Thoroughly remove all flux remaining on the casting.

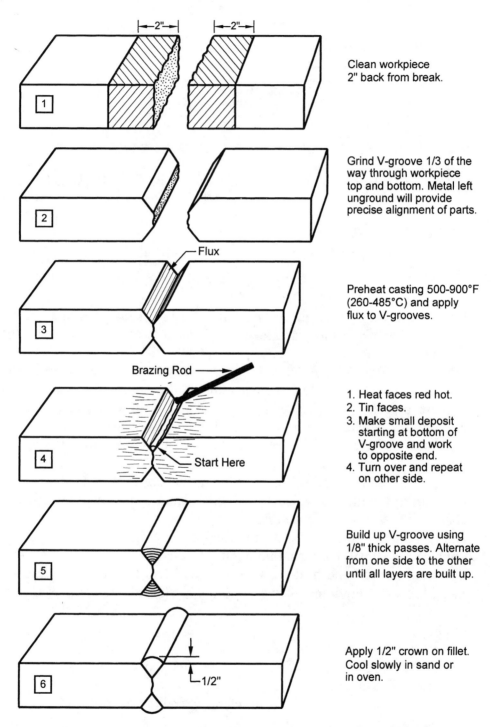

Figure 6–122. Steps in braze welding a broken casting.

Chapter 7

Strength of Materials

A problem well stated is a problem half solved.
—Charles F. Kettering

Introduction

Strength of materials studies the distribution of forces *inside* structures, how structures deform under load, and how they fail. Understanding these forces and their distribution allows engineers to design and size load-carrying beams, shafts, and columns with economy and safety.

Engineering schools take a semester to study strength of materials, so this chapter is but a brief overview of the main issues. It provides exposure to stress, strain, modulus of elasticity, tensile testing, moment of inertia, and metal fatigue. It examines how stress is distributed in rectangular beams and cylinders, and how some shapes of beams and shafts can carry loads more efficiently. It also looks at how beam span affects beam deflection.

Mechanics is a mature science and its theoretical basis well developed, but before the development of mechanics, experience—especially previous structural failures—guided designers. To avoid failures, most mechanical and civil engineers massively overdesigned their projects using many times too much material. The widespread development of the industrial revolution brought the need for roads and bridges, and with that came economic considerations and the need for mathematical analysis to predict how much material was really needed.

Section I – Mechanical Property Definitions

What is the definition of each of the following terms?
- *Brittleness* is that property of a material having no plastic deformation before failure and is the opposite of ductility and malleability. Brittle materials have no yield point and have a rupture strength and ultimate strength that are approximately equal. Cast iron, ceramics, concrete, and stone are brittle and relatively weak in tension compared with most other engineering materials. Brittle materials are usually tested in compression.
- *Ductility* is that property of a material permitting it to undergo considerable plastic extension *under tensile load* before fracture. Gold, aluminum, and copper are ductile materials and easily drawn into wire.
- *Elasticity* is that property of a material that allows it to regain its original dimensions when the deforming load is removed. Steel is elastic up to its elastic limit.
- *Malleability* is that property of a material that allows it to undergo considerable plastic deformation *under compression* before rupture. Rolling and hammering operations are used on malleable materials. Gold is both malleable and ductile.
- *Plastic deformation* occurs when a material has been stressed above its elastic limit and retains some or all of the deformation it had under load after the load has been removed.
- *Resilience* is that property of a material that allows it to withstand a large, suddenly applied stress without exceeding its elastic limit.
- *Stiffness* is that property of a material that allows it to sustain high stress without great strain. Stiffness is measured by a material's modulus of elasticity, E. Steel with an E of 30,000,000 psi is a stiff engineering material; wood with an E of 1,000,000 psi is not.
- *Strength* of a material is the greatest stress that the material can sustain without failure. While it may be defined by the proportional limit, yield strength, rupture strength, or ultimate strength, no single parameter can define this property as the material's behavior varies with the kind of stress and the way it is applied. Strength is a general term.
- *Toughness* is that property of a material that allows it to sustain rapidly applied or shock loads.
- *Yield Point,* or *Yield Stress* is that level of stress in metals measured in psi, or MPa above which the metal remains stretched after stress is removed. Plastic deformation has occurred.

Section II – Stress & Strain

Stress

What is Hooke's law?

Robert Hooke (1635-1703) found that the deformation of an elastic body is directly proportional to the magnitude of the applied force, provided the elastic limit is not exceeded, Figure 7–1. Put another way: the force on and deformation of a material are proportional (linear) until it reaches a very high stress level where the sample material becomes permanently deformed. Hooke also found that solid materials like metals *do* actually change their shape under load, but to a small degree—usually a fraction of one percent.

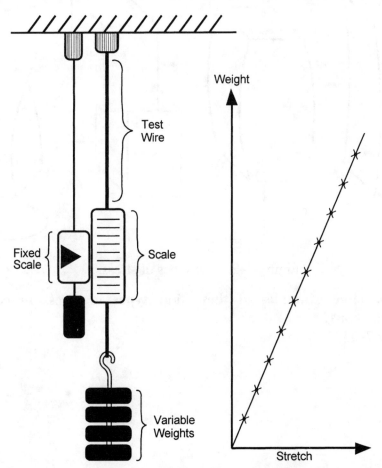

Figure 7–1. Stretch of a wire is directly proportional to the weight on it.

What are the four types of stress?
- *Tension* which tends to lengthen the material.
- *Compression* which tends to shorten the material.
- *Shear* which acts to offset the material in a plane parallel to the stress.
- *Torsion* which tends to twist the ends of the material in opposite directions.

All stress, however complex, can be described by a combination of two or more of these basic types, Figure 7–2.

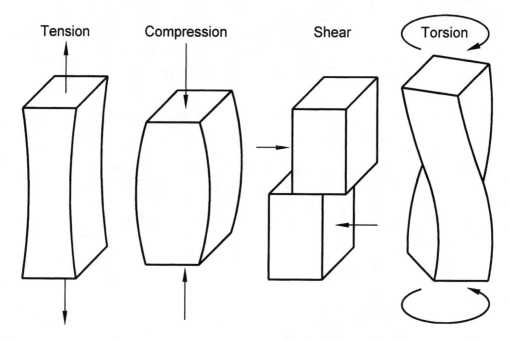

Figure 7–2. Four types of stress.

What are some examples of these four types of stress in common mechanical items?
See Figure 7–3.

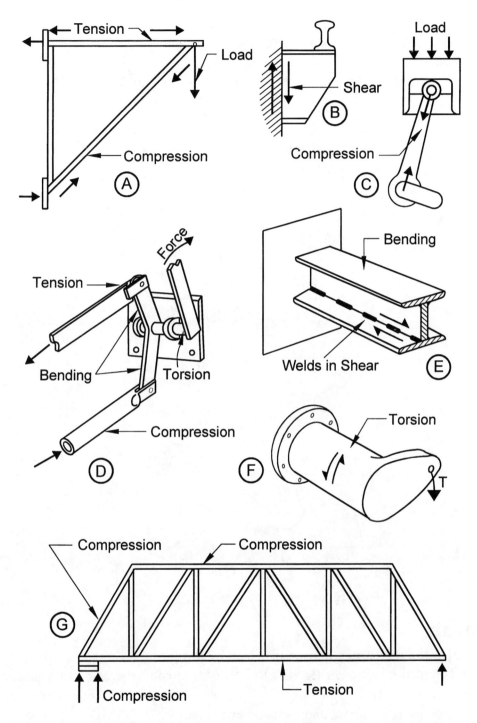

Figure 7–3. Examples of the four types of stress in practical mechanisms.

How is stress calculated?
Stress is the force divided by the area on which it acts and is measured in pounds/inch², also written *psi*. In the metric system kiloPascals (kPa), or megaPascals (MPa) are the units of measure. See Figure 7–4.

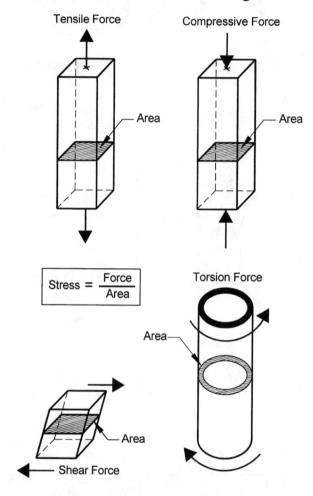

Figure 7–4. How stress is calculated.

Strain

What is the difference between *stress* and *strain*?
The application of *stress* (loading) to a material causes *strain* (deformation). Strain is the ratio of the dimension of the test material before the stress application to its *change* in dimension after loading.

What units does strain have?
Because it is a ratio, strain is dimensionless. See Figure 7–5.

WELDING FABRICATION & REPAIR 257

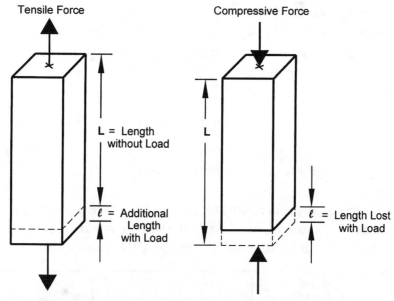

For Tensile and Compressive Forces:

$$\text{Strain} = \frac{\text{Change in Length}}{\text{Initial Length}} = \frac{\ell}{L} \text{ (Dimensionless)}$$

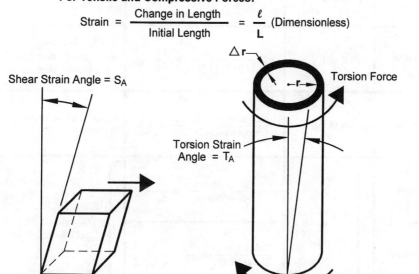

For Shear Force:
  Strain = S_A = Shear Strain Angle in Radians (Dimensionless)
For Torsion Force:
  Strain = T_A = Torsion Strain Angle in Radians (Dimensionless)

Figure 7–5. How strain is calculated.

Modulus of Elasticity

What is the relationship between stress and strain?
Provided the material is stressed within its elastic range, stress and strain are proportional. They are directly related by a number called the *modulus of elasticity*:

$$\text{Stress} = (\text{modulus of elasticity}) \times \text{Strain}$$

There are several types of stress and strain. Does one modulus of elasticity number relate to all these types of stress and strain for a given material?
No, each type of stress is related to its respective type of strain by the modulus of elasticity for the particular type of *stress-strain pair*. These stress-strain pairs and their modulus of elasticity are shown in Table 7–1.

Type Stress	Symbol for Stress/Strain Ratio (lb/in^2 or kPa)	Formula	Alternate Names
Tensile	E or E_Y	$\dfrac{\text{Tensile Stress}}{\text{Tensile Strain}} = \dfrac{FL}{A\ell}$	Young's Modulus Modulus of Elasticity in Tension Longitudinal Modulus of Elasticity
Compressive	E or E_C	$\dfrac{\text{Compression Stress}}{\text{Compression Strain}} = \dfrac{FL}{A\ell}$	Modulus of Elasticity in Compression
Shear	G or E_S	$\dfrac{\text{Shear Stress}}{\text{Shear Strain}} = \dfrac{F_S}{S_A}$	Shear Modulus Modulus of Elasticity in Shear Coefficient of Rigidity
Torsion*	G or E_S	$\dfrac{\text{Shear Stress}}{\text{Shear Strain}} = \dfrac{F_T}{T_A} = \dfrac{F_T}{2\pi r \Delta r}$	

*Although caused by a torsional force, shear stress, and strain result.

Table 7–1. Modulus of elasticity.

The *modulus of elasticity in tension*, also called *Young's Modulus*, is often symbolized by E or E_Y, and is used much more frequently than the modulus of elasticity in compression or in shear. When someone says "modulus of elasticity" without further qualification they usually mean *modulus of elasticity in tension*.

WELDING FABRICATION & REPAIR

Although torsion is classified as one of the four basic types of stress, it produces shear stress and shear strain when applied. There is therefore no separate modulus of elasticity for torsion force.

What are the measurement units of the modulus of elasticity?
The units of all modulus of elasticity are the same as stress: *psi* in English units, or *kPa* or *MPa* in metric units.

How can we use the modulus of elasticity to solve a practical problem?
Here is an example of how we can calculate the change in the length of a rod or wire under load. Given a steel rod of 50 foot length and 0.25 inch diameter, determine the stress in the rod and how much it will stretch when loaded with 500 pounds, Figure 7–6. Assume $E_Y = 30$ million psi.

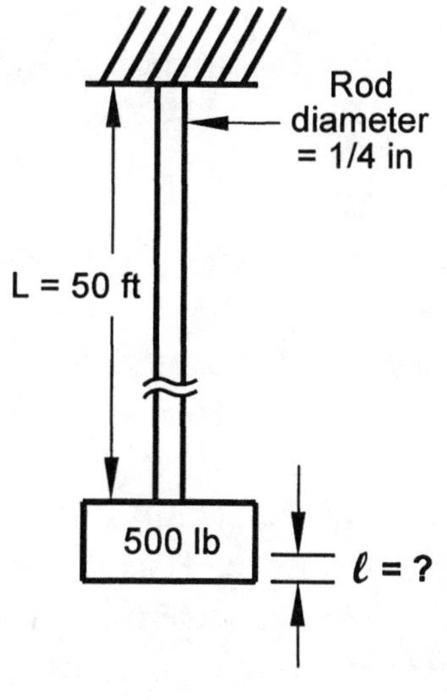

$$\text{Stress in rod:} \frac{\text{Force}}{\text{Area}} = \frac{500}{\pi \left(\frac{1}{2} \cdot \frac{1}{4}\right)^2} = 10186 \text{ lb/in}^2$$

$$\text{Stress in rod:} E_Y = \frac{FL}{A\ell}$$

$$\text{then } \ell = \frac{FL}{A E_Y}$$

$$= \frac{500 \text{ lb} \times 50 \text{ft} \times 12 \text{ in/ft}}{\pi \left(\frac{1}{2} \cdot \frac{1}{4}\text{in}\right)^2 \times 30 \times 10^6 \text{ lb/in}^2}$$

$$= 0.204 \text{ in}$$

Figure 7–6. Steel rod stretching under load.

Given either stress or strain, we can determine the other using this formula.

Is there another way to visualize the meaning of the modulus of elasticity?
Yes, the numerical value of E is the stress required to stretch a specimen to *twice* its original length (2L), if it did not break before then. In fact, most metals will not survive much more than one percent strain, but it is still an important concept. Table 7–2 shows the modulus of elasticity for common metals.

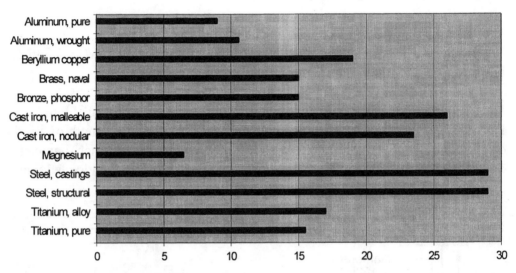

Table 7–2. Modulus of elasticity, E, for common metals $\times$ 10^6 psi.

How does the modulus of elasticity in *compression* differ from the modulus of elasticity in *tension*?

Technically the modulus of elasticity in compression and the modulus of elasticity in tension must each be measured under the conditions of compression and tension, respectively. However, the modulus of elasticity in compression and the modulus of elasticity in tension are the same value *for metals* under the stress levels seen in engineering applications. Tensile tests are much more convenient than compression tests, so we determine the modulus of elasticity in tension and use this value for the modulus of elasticity in both tension *and* compression for metals.

Non-metallic materials, like brick, stone, and concrete are weak in tension and strong in compression and cannot easily be tested in tension. The modulus of elasticity in compression for these materials must be determined with a compression test.

Why is the *modulus of elasticity* of a material important?
There are many reasons:
- E provides an accurate measure of the stiffness or rigidity of a material and allows us to accurately compare one material's stiffness with another's, Figure 7–7.

- *E* always appears in engineering formulas for determining stress and deflection in beams and columns, so knowing its value is essential for these calculations.

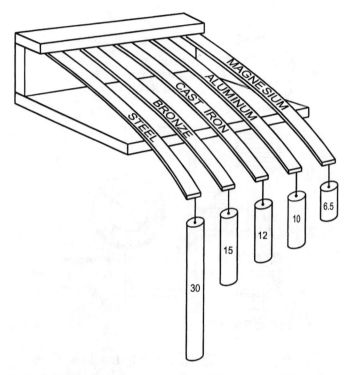

Figure 7–7. Weight required to produce the same deflection of these identically sized bars is proportional to the modulus of elasticity of its metal.

Section III – Stress-Strain Curves

Tensile Testing

How is *tensile testing* performed?

Tensile testing is done by pulling apart a test specimen in a testing machine while measuring both stress and strain. Figure 7–8 shows a typical test specimen and the measurements made on it as increasing tension force is applied. Figure 7–9 shows a typical stress vs. strain curve for carbon steel. While this curve could have been determined by taking a succession of readings from the tensile testing machine's gauges and plotting the points, today electronic sensors attached to the testing machine capture this data and send it to a plotter. From this curve, we can quickly determine yield strength and ultimate tensile strength, and so evaluate materials for a particular application. Such tests also serve to confirm that steel has the yield and ultimate tensile strength claimed by the manufacturer and is properly marked.

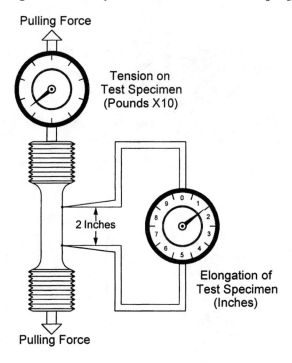

Figure 7–8. Tensile testing machine.

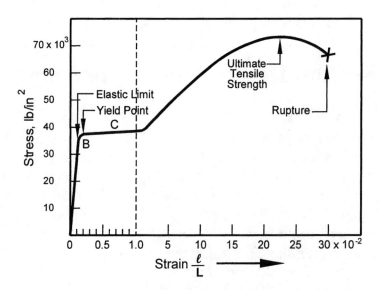

Figure 7–9. Stress-strain curve from testing machine for carbon steel.

What information does a stress-strain curve provide about a material?
- *Modulus of Elasticity*—The slope or steepness of the straight-line portion of the stress-strain curve, the portion below the yield point, gives the modulus of elasticity in tension. The steeper this slope, the stiffer the metal.
- *Elastic Limit*—The greatest stress the material can withstand without permanent elongation when all load has been removed from the specimen. Any strains developed up to the elastic limit are both small and reversible.
- *Proportional Limit*—The greatest stress a material can withstand without deviation from the straight-line proportionality between stress and strain. For practical purposes, the elastic limit and the proportional limit occur at the same stress.
- *Yield Strength*—The stress at the uppermost point on the straight-line portion of the curve. Stress imposed on the sample below this level produces no permanent lengthening and stress can vary from zero up to the yield strength. Stress above yield strength causes permanent lengthening. Engineers design structures so that stress in the components remains much lower than the yield strength. Exceptions to this would be items such as pull-tab openings on soda cans that are *designed* to fail when used.
- *Ultimate Tensile Strength (UTS)* or *Tensile Strength*—This is the maximum stress the sample will withstand before failure. In ductile metals, the metal has already been permanently deformed. While this is never a desirable stress-strain level in structures, it provides a measure of the ductility of a material.

Why is the *tensile testing* of a material important?
Since structural designs must hold stress well below a material's yield strength to avoid permanent stretching, knowing yield strength is essential. Performing a tensile or pull test is a quick, accurate, and convenient way to evaluate a material's strength.

How do stress-strain curves compare for a variety of engineering metals?
Figure 7–10 show the stress-strain curves of several engineering metals. These curves are useful because they show the yield and ultimate tensile strength. They also allow us to compare the ductility of these metals.

If we expanded the strain scale at its lowest end, we could measure the modulus of elasticity in tension for each material and compare their stiffnesses.

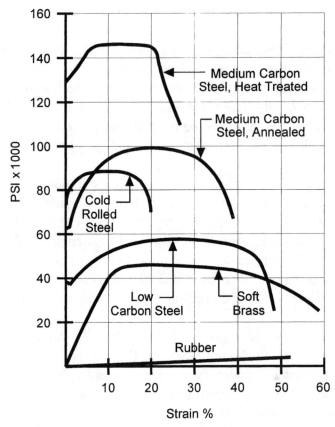

Figure 7–10. Stress-strain curves for a variety of engineering materials.

Are there other ways to determine the tensile strength of a material?
Although the most accurate tensile testing is done by pulling a specimen apart in a tensile testing machine, such testing destroys the sample, and cannot be done in the field. We can get a good estimate of a material's tensile strength by performing a hardness test with a portable testing machine and converting the hardness number to tensile strength using a conversion chart. This approach is low-cost, fast, and non-destructive.

An even lower cost instrument to make a rough, non-destructive test on weld hardness in the field is to use the simple instrument shown in Figure 7–11. This device drops a ball bearing onto the test weld and measures the rebound on a scale. The amount of rebound indicates sample hardness and tensile strength.

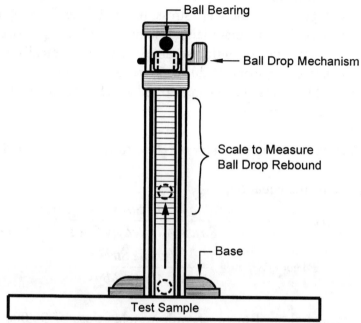

Figure 7–11. Ball-drop portable hardness tester.

Is there a difference between the *strength of a structure* and *strength of a material*?
Yes, the strength of a structure is its load-carrying capacity when put together in a particular way with a given set of materials. The strength of a structure is given in pounds or kilograms. Exceeding this load could damage the structure. The strength of a material is a physical property that is not dependent on the way the material is incorporated into a structure.

Is there a typical upper limit of strain for engineering materials?
Yes, typically about one percent.

Allowable Stress

What is allowable stress and who determines what it should be?
Allowable stress is the maximum stress, in psi or MPa, permitted in a structural member. The idea is to set the maximum allowable stress level at such a low level that the structural member will always remain in the elastic zone under all conditions.

This stress level is set by building codes and engineering societies and is the cumulative result of history and experience. The maximum allowable stress is typically 20 to 60 percent of the yield stress, but will depend on many factors. To provide for variations in materials, customer-overloading, uncertainty in calculating the stress, and the unforeseeable, engineers typically design and size stressed materials to work at 20 to 50 percent of their yield strengths. Setting a low working stress level can also reduce chances of crack formation and fatigue as discussed below.

How does maximum allowable stress relate to the figure called *factor of safety*?
They are related by the equation:

$$\text{Factor of Safety} = \frac{\text{Yield Stress}}{\text{Stress Allowed by Code}}, \text{ or}$$

$$\text{Factor of Safety} = \frac{\text{Tensile Stress}}{\text{Stress Allowed by Code}}$$

If we had a structural steel member with a yield stress of 36,000 psi and a maximum allowable design stress of 24,000 psi, we would have:

$$\text{Factor of Safety} = 36,000/24,000 = 1.5$$

This is the same as saying that this structural member has 50% reserve strength against yielding.

Compressive Strength

What about testing *compressive strength* in metals?
Compressive strength, the ability of a metal to resist a gradually applied squeezing force, is not usually important for metals because engineering materials have at least as much strength in compression as in tension. In fact, Young's Modulus in tension can equally be used as the modulus of compression. At very high stress levels, metals will experience permanent deformation just as in tension, and more brittle building materials like brick and concrete will fracture. However, deformation and failure will generally be at a much higher stress level than the material's tensile strength.

Measuring compressive strength is more complicated than measuring tensile strength because in ductile materials like steel, test material is actually moved or displaced before failure. Because of this, the sample shape and size has a big influence on test results.

Section IV – Beams

Stress Distribution in Beams

What are the three types of stress in a rectangular beam supported at its ends and loaded at its middle?

- *Bending stress* produces compression on the loaded side of the beam, tension on the other, Figure 7–12.

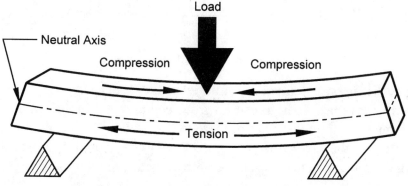

Figure 7–12. Load causes bending stress that produces compression on the top and tension on the bottom of the beam.

- *Horizontal shear stress* results when horizontal layers of the beam attempt to slide past each other due to the varying levels of compression and tension between the upper and lower beam layers, Figure 7–13.

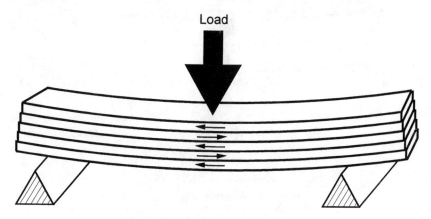

Figure 7–13. Horizontal shear stress caused by different levels of compression or tension in adjacent beam layers.

- *Vertical shear stress* results from the load pressing down on one side of the beam and the other side being pushed up by the supports, Figure 7–14.

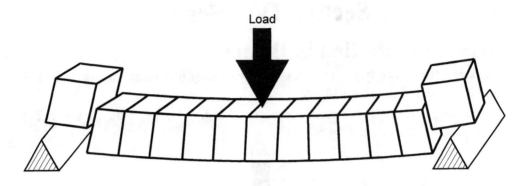

Figure 7–14. Vertical shear stress.

Where is the beam's neutral axis and why does it occur?
As beam stress makes the transition from compression along the top of the beam to tension along the bottom, there is a line along the beam where stress is zero; this is the *neutral axis*, Figure 7–15. The neutral axis lies midway between the top and bottom of beams with vertical symmetry.

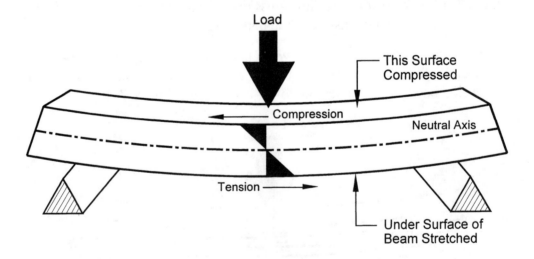

Figure 7–15. Distribution of compressive and tensile stress above and below the beam's neutral axis.

What can we conclude from studying the stress distribution within the beam in Figure 7–15?

- Beam material most distant from the neutral axis is under the greatest stress, compression on the top and tension on the bottom.
- Beam material closest to the neutral axis is under little stress and so is carrying very little of the beam load and contributes little to beam stiffness.
- Because beam material near the neutral axis is doing little to contribute to beam stiffness, cutting a hole in the beam on or near the neutral axis for pipe, or conduit weakens the beam very little. Centering beam holes on the neutral axis is good practice, and conversely, cutting through the beam material where it is under great stress—along its top or bottom—will weaken the beam, Figure 7–16.

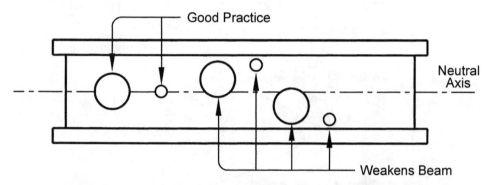

Figure 7–16. Good and bad practice for drilling through an I-beam.

- If the beam material near the neutral axis does not contribute much to beam stiffness and load carrying ability, then we can remove a lot of this material without much effect on the beam stiffness, or we can redesign and reshape the beam and make it stronger using the same quantity of material.

When all the stresses inside a beam are combined, what do the stress patterns look like?
Figure 7–17 shows the *flow* of tensile and compressive forces within a beam, also called the *stress trajectories*. Only the direction of the forces are shown, not the magnitude.

We calculate forces inside a beam—tension and compression, vertical shear and horizontal shear—individually for two reasons: First it simplifies the analysis and second, each type of force has its own type effect on the beam material. And by knowing the size of each type force, we can predict beam behavior.

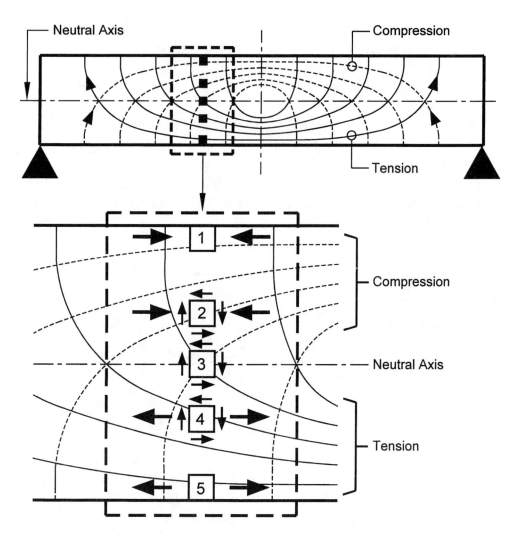

Figure 7–17. Stress trajectories in a beam.

What can we conclude from studying the stress trajectories of the beam in Figure 7–17?

Each of the numbered boxes in the detail view of Figure 7–17 represents a tiny element of beam material.

- On the top and bottom of the beam only bending forces exist. These bending forces create compression in box 1, and tension in box 5.
- In box 3 at the middle of the beam on the neutral axis, only shear stress exists and runs at a 45° angle to the neutral axis. The crossing of tension and compression trajectories on the neutral axis at 45° shows this. No compression or tension forces exist on the neutral axis.

- The beam material elements in box 2 experience compression force and shear force, while beam material elements in box 4 experience tension and shear force.

How can the conclusions from Figure 7–17 be applied in practice?
By knowing where and how a beam is stressed, we can redesign it to be stronger and lighter. This truss is a good example. The shape of trusses mirrors the location and direction of stress in a beam. Designers place a strong top member to carry the compressive stress, greatest along the top of the beam. Because it takes a smaller beam to carry the same tensile stress, a smaller, lighter member is adequate on the truss bottom. The light truss rods running at a 45° angle *parallel the direction of beam shear forces* and carry the relatively light shear loads. This is an efficient beam design as it eliminates beam material where it is not heavily stressed, Figure 7–18. Trusses can span greater distances with less material and weight than a solid beam, because their load-carrying members neatly match the stresses imposed.

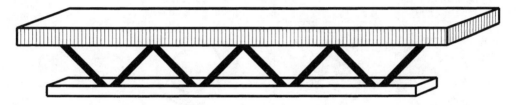

Figure 7–18. Truss shape mirrors beam stress flows.

Moment of Inertia, *I*

What is *moment of inertia* and why is it important in the study of strength of materials?
Moment of inertia, also called the *second moment of inertia*, is a number that provides a measure of beam stiffness based on its size and shape. Moment of inertia has units of inches^4 or mm^4.

Together moment of inertia and modulus of elasticity tell us all we need to know to compare the stiffness of one beam with another and to calculate beam stress and deflection under load. *Because beam size and shape determine moment of inertia and beam material determines modulus of elasticity, these two parameters neatly separate the effect of beam shape and beam material.* All beams of the same size and shape have the same moment of inertia regardless of whether they are wood or steel. Similarly, all beams of the same material have the same modulus of elasticity regardless of their size and shape.

How is moment of inertia determined?
There are a variety of ways:
- For I-beams and other rolled shapes, we can look them up in tables that list their dimensions, weight/length, and moment of inertia. These tables are found in mechanical engineering handbooks, or data sheets from steel companies.
- For geometric shapes, engineering handbooks list formulas for calculating *I*. See Table 7–3.

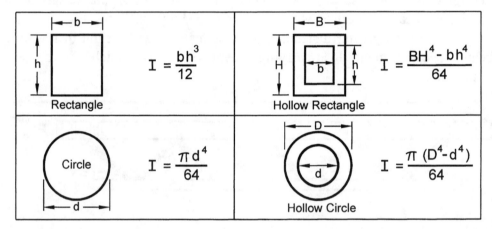

Table 7–3. Formulas for calculating moment of inertia for geometric shapes.

For the three beams shown in Figure 7–19, is one shape more efficient than another in providing stiffness?
We can compare beam stiffness by comparing moments of inertia. All three shapes have the same cross-sectional area, weight/foot of length, and so total amount of material.

Welding Fabrication & Repair

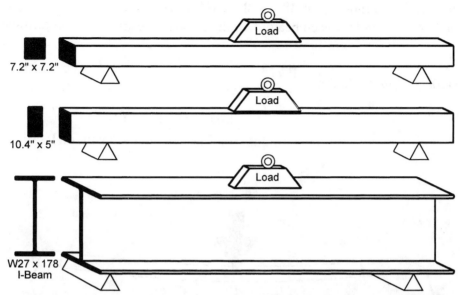

Figure 7–19. Three beams of the same material and cross-sectional area.

We calculate the moment of inertia of the square and rectangular beams using the formula in Table 7–3 (rectangle, top left corner), and look up the moment of inertia for the wide flange I-beam in beam tables. These results are shown in Table 7–4.

	Dimensions (inches)	Area (inches^2)	Moment of Inertia (inches^4)	Stiffness Improvement over Square Beam
Square Beam	7.2 × 7.2	52	224	1.0
Rectangular Beam	10.4 × 5	52	468	2.1
Wide Flange I-Beam	W27 × 178	53	6990*	31.2

* From AISC WF Beam Tables

Table 7–4. Comparison of moments of inertia for three beams in Figure 7–19.

The conclusion is that by reshaping the same amount of material, the beam can be made 31 times stiffer. This remarkable result is true whether the three beams are steel, plastic, or wood. The moment of inertia is solely determined by the beam's shape, not its material. Shape does count.

Beam Deflection

If each of the beams in Figure 7–19 are steel and carry a 5000-pound load at its center across a span of 40 feet, what is the deflection at the center of

each beam? Assume $E_{Steel} = 30 \times 10^6$. See Figure 7–20.
Table 7–4 lists the *I* for each beam. Use the standard beam deflection formula:

$$\text{Deflection @ Midpoint} = \frac{P \times L^3}{E \times I}$$

$$= \frac{\text{Load} \times (\text{Length of Span})^3}{(\text{Modulus of Elasticity}) \times (\text{Moment of Inertia})}$$

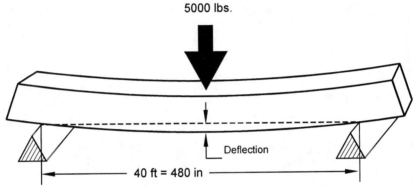

Figure 7–20. Calculation of beam deflection.

Table 7–5 shows that even though all beams contain the *same amount* of material, an I-beam has the least deflection and is most efficient. The deflection is so small a much smaller I-beam would be suitable.

Beam	Deflection (inches)
Square Beam	1.7
Rectangular Beam	0.8
W27 × 178 Wide-Flange I-Beam	0.05

Table 7–5. Calculated beam deflections.

How much beam deflection is considered acceptable?
While a particular building code or structure may have special requirements, many codes limit deflection to 1/360 of the beam's span. This much sag is not easily noticeable and will not lead to cracked plaster and sticking doors.

Is there a general rule of thumb to quickly determine if an I-beam will produce a small enough deflection?
Yes, if the length of the span divided by the depth of the I-beam is under 23, deflection is not likely to be a problem. In the case of the W27 × 178 I-beam (27 inches deep and 178 pounds/foot) in Figure 7–19, the span is 40 feet (480 inches) and the beam depth is 27 inches. Using this rule and dividing 480/27, we get 17.7. Because this result is less than 23, the deflection is acceptable.

When selecting the size of a beam, which is the deciding factor, deflection or load-carrying ability?
Deflection, because most beams reach their maximum allowable deflection long before they reach their maximum bending or shear stress.

Beam Deflection with Length of Span

For a beam supported at its ends and carrying a load at its center, what effect does changing the length of the span have on beam deflection?
Because the applicable beam deflection formula (above Figure 7–20) has the length of the span cubed:
- Halving the span length reduces deflection to 1/8 as much.
- Doubling the span length increases deflection 8 times.

Reinforcing Concrete

What is the range of concrete's compressive and tensile strength?
Depending on the mix and age, concrete will have a compressive strength between 2500 and 7600 psi. *Although concrete has a tensile strength of about 200 psi, engineers assume it has none.* A beam of only concrete cannot carry much load and is likely to crack of its own weight, Figure 7–21 (top).

If concrete has no tensile strength, how do concrete beams carry a load?
Steel reinforcement bars, called *rebar*, are cast *inside* the concrete. Rebar on the bottom of the beam carries the tensile load, so the concrete does not have to. Adding this rebar is a step in the right direction, but the shear stress which is greatest near the beam supports causes cracks to occur as shown in Figure 7–21 (middle). Bending the rebar so it runs across the line of shearing to carry the tensile part of the shear load is the solution, Figure 7–21 (bottom). Combining concrete, with its high compressive strength, and steel, with its high tensile strength, takes advantage of the best properties of both materials. The result is a fireproof, rustproof beam.

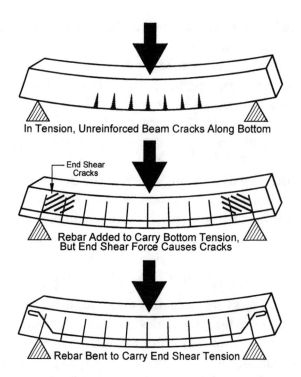

Figure 7–21. Reinforcement of concrete beams.

What is *prestressed* concrete, how is it made and what are its advantages?
Prestressed concrete has steel rods or cables *inside* it to put the concrete under high compression *before* the concrete is put under external load. This is done so that when additional tension is added by the load, the beam is still in compression.

There are several ways to produce prestressed concrete beams; here is one of them. First, using wood or steel forms, a concrete beam is poured with PVC pipes cast inside near the bottom face of the beam, Figure 7–22 (a). Beams that fit on a truck are factory-made, larger ones are cast in place. The PCV pipes provides channels for threading steel rods (or steel cables), Figure 7–22 (b). When the concrete hardens, hydraulic puller jacks tension the rods, and the rods are secured so they remain under tension to compress the beam along its lower part, Figure 7–22 (c). Load on the beam when in service reduces the compression on the beam somewhat, but there is enough tension in the rods to keep the beam in tension, Figure 7–22 (d).

Prestressing concrete with rods or cables makes concrete beams significantly stronger than just adding rebar. Prestressing is widely used for bridges today.

Welding Fabrication & Repair

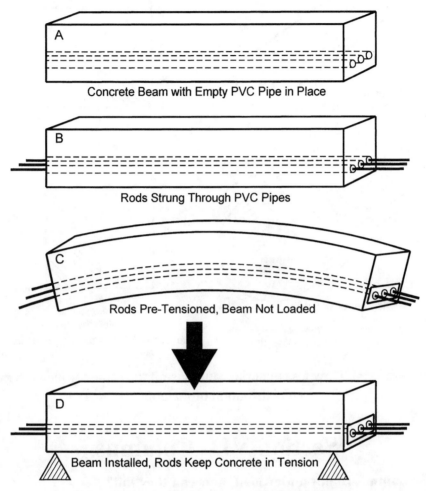

Figure 7–22. Prestressing a concrete beam.

Section V – Shafts

Stress Distribution in Shafts

How can we redistribute the material in a shaft under torque to transmit the same load with less weight?

We need to reshape the shaft so more of the material is under greater stress. The interior of the shaft carries little stress and only the outermost portion of the shaft is actually "working" for us, so the solution is to remove material from the center of the shaft and go to a larger diameter *hollow* shaft. See Figure 7–23.

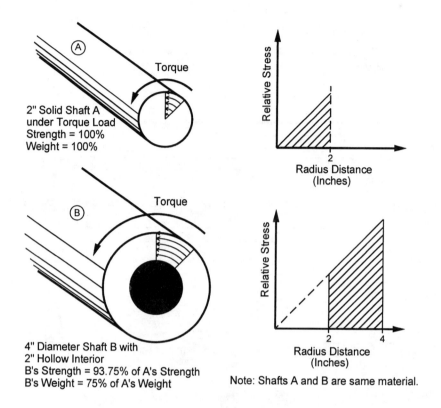

Figure 7–23. How less material can transmit the same torque with a redistribution of material.

Section VI – Columns

When columns are put under load, how can they fail?
Short columns fail under compression; they sustain such heavy loads that the column material fails by fracturing or deforming under compression. Long columns fail by *bending,* sometimes called buckling at much lower loads.

What effect does the way the column is held in place have on its ability to support a load?
On long thin columns, Figure 7–24, the "end conditions" determine how much load a column can support. Table 7–6 shows the number of times stronger a column pinned at both ends is over a column pinned at only one end.

WELDING FABRICATION & REPAIR

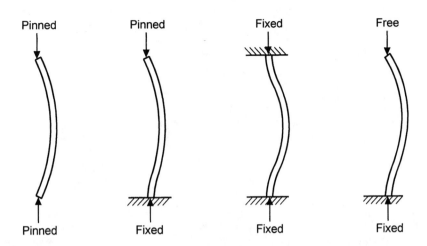

Figure 7–24. Different end conditions determine the columns' safe load.

Column End Conditions	Times Stronger Than Pinned End Column
Pinned/Pinned End Column	1
Fixed/Pinned End Column	2
Fixed/Fixed End Column	4
Fixed/Free End Column	1/4

Pinned = column can flex at ends but not move sideways.
Fixed = column is cast in concrete or otherwise rigidly captured.
Free = able to move, like the top of a flagpole.

Table 7–6. Comparison of beam safe loads with three end conditions.

Section VIII – Other Stresses

Sudden Stress

What happens to the stress in a structure when a load is suddenly applied?

Figure 7–25 shows that engineers must plan for *twice* the stress when a given load is suddenly applied. Stress from loads actually dropped on a structure will be significantly greater and require detailed analysis.

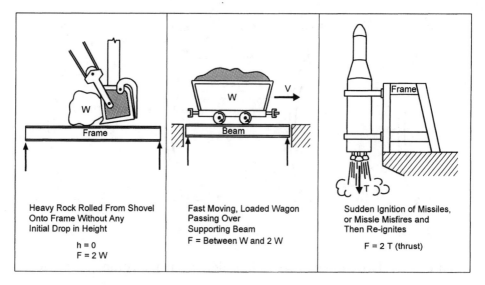

Figure 7–25. Forces applied suddenly.

Section IX – Metal Fatigue

What is *metal fatigue*?
Metal fatigue is the development and gradual growth of a crack, which suddenly fractures a component after repeated cyclic stress. Metal fatigue is a phenomenon of crystalline materials.

At what stress levels do metal fatigue failures occur?
Fatigue-induced failures often occur at one-half to one-quarter of the maximum elastic load the metal can sustain under constant stress. In extreme fatigue cases, as little as one-eighth the maximum elastic load causes failure after enough load cycles.

Why is metal fatigue important?
Even if a metal part is properly sized to carry its anticipated maximum loads, fatigue failure can eventually occur, causing inconvenience, property damage, and, even loss of life.

Metal fatigue first became important with the advent of the steam engine in the eighteenth century. Although engineers had enough understanding of physics and materials to design steam engine components to handle the anticipated *static* loads, the cyclic loads developed by the steam engine introduced these early engineers to metal fatigue, sometimes with catastrophic results. Even at two hundred revolutions per minute, a machine will see over 100 million load cycles after a year. Before steam power, most metal structures were bridges or machines subject to few load cycles. Suddenly metal fatigue was an issue.

Today 90% of all 'in use' failures are fatigue failures, not stress overload failures. Proper design prevents fatigue failures. Replacing critical parts after a set number of load cycles or hours of service also prevents in service failures. Careful periodic inspections can spot signs of fatigue failure *before* it occurs.

How does fatigue failure begin?
Despite the fact that the final result of fatigue failure is a sudden and complete breakage of the part, fatigue failure is a *progressive* failure. It may take a very long time after beginning to reach complete failure.

Fatigue cracks begin where a geometric irregularity in the part such as a hole, crack, scratch, sharp corner, or metallic defect creates a *stress concentration*. Often this increase in stress occurs in a very small area. Although the average unit stress across the entire part cross-section may be well below the metal's yield point, a non-uniform distribution of stresses causes stress to exceed the yield point in some minute area and causes plastic deformation. This eventually leads to a small crack. This crack aggravates the already non-uniform stress distribution and leads to further plastic deformation. With repeated load cycles, the crack continues to grow until the cross-section of the part can no longer sustain the load and final and catastrophic failure occurs. The origin of the crack is on or near the part's surface because that is where stresses are greatest.

Each load cycle performs work to break the bonds between metal atoms and enlarge the crack. In general, these cracks split the metal's grains, rather than preferentially running along grain boundaries. This *intragrain* fracturing creates a concentric ring pattern radiating from the failure's origin called "beach marks" after the pattern left by the sea as it retreats from a sandy beach, Figure 7–26 (left). When the crack become large enough that the remaining metal cannot sustain the stress, catastrophic failure occurs. This failure is *intergranular*, or between the grains, and leaves a different kind of pattern between the failed surfaces. The beach mark pattern is caused by uneven crack growth and the rubbing together of the sides of the crack as it develops.

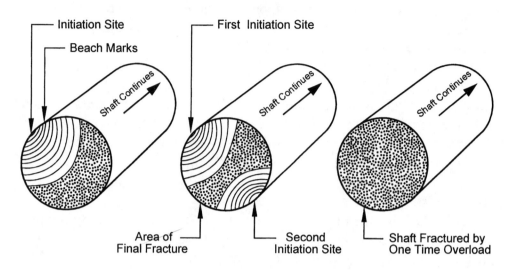

Figure 7–26. End view of a fractured shaft: Typical beach marks from fatigue crack growth (left), growth of a second fatigue crack with two sets of beach marks (center), and shaft failure by one-time overload (right).

What can examining the fracture interfaces tell us?
Because the semicircular shaped beaching pattern inside the crack radiates from the crack's initiation point, we can determine its point of origin. Sometime after one crack initiates, another crack develops on the opposite side of the part. This second initiation site will also be visible, Figure 7–26 (center). Another clue to understanding what happened to cause fatigue failure is that the plane of fatigue fracture is always perpendicular to the direction of maximum stress.

By examining the ratio of the beaching pattern area versus the intergranular final fracture area, we can see how much margin of safety the part had initially. The greater the area of beaching, the more material the crack had to work through before failure. A failed part with a small beaching area had a smaller margin of safety than a similar part with a larger beaching area.

A failed part without beaching indicates a one-time overload, not a fatigue failure, Figure 7–26 (right).

What causes stress concentrations?
Figure 7–27 (top) shows a metal bar under stress with no defects or irregularities. The stress trajectories shown by the dotted lines show stress

WELDING FABRICATION & REPAIR

divides evenly across the part. Average stress is the maximum stress. Figure 7–27 (bottom) shows a similar part with a notch also under stress. The stress trajectories represented by the dotted lines "bunches up" around the tip of the notch. These are areas of stress concentration. Peak stress is much higher than average stress. Stress concentration in this area will cause a crack as the part is load cycled and the crack will grow.

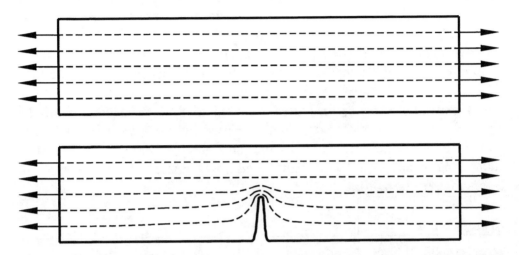

Figure 7–27. Stress trajectories in a flat bar under tension: Bar without defects (upper), and bar with crack entering its side creating stress concentration at crack tip (lower).

In the simple case of a round hole drilled through a metal bar, Figure 7–28, how much of a stress increase does adding the hole cause?
Table 7–7.

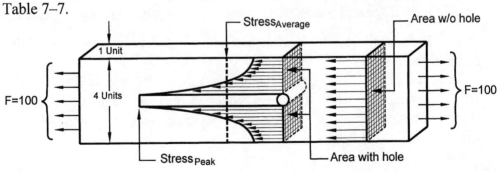

Figure 7–28. Stress increases in a rectangular metal bar under tension caused by the addition of a hole.

Two factors affect stress levels in the bar. Adding a hole to the bar:

- Reduces the area of the bar on each side of the hole carrying tension force from 4 area units to 3 area units, so Stress$_{Average}$ increases from 25 to 33.3 units, Table 7–7.
- Disrupts the smooth flow of stress through the metal bar, and produces stress concentrations at the edges of the hole. Unlike sections of the bar without holes, the average stress is no longer peak stress. In fact, the stress around a hole like the one in Figure 7–28 is about 2.5 times the average stress. This is called the *stress concentration factor* or *SCF* for short. Figure 7–28 and Table 7–7 shows how the SCF increases Stress$_{Peak}$ to 82.5 units.

In this example, Stress$_{Peak}$ increased from 25 to 82.5 units because of the addition of a hole. Stress concentrations caused by holes must be taken into account by designers for both static and fatigue strength. Mechanical engineering handbooks have charts used to estimate stress concentration factors for various part layouts.

Bar Cross-section	Area	Force	Stress$_{Average}$	Stress$_{Peak}$
No Hole	4	100	$\dfrac{F}{A} = \dfrac{100}{4} = 25$	Stress$_{Average}$=Stress$_{Peak}$= 25
With Hole	3	100	$\dfrac{F}{A} = \dfrac{100}{3} = 33$	Stress$_{Peak}$ = SCF × Stress$_{Average}$ = 2.5 × 33.3 = 82.5

All units are relative.

Table 7–7. Average and peak stress for cross-sections of a bar under tension.

How much does the crack or other irregularity increase stress above the average stress?

Usually holes, fillets, and edges will create stress concentrations from 1.2 to 3 times the average stress at the cross-section. The increase of stress above average stress depends on the shape of the specific workpiece, and the size, shape and location of the stress raiser.

Holes, interior fillets, exterior fillets, sharp corners, or steps also behave like cracks—they serve to create a point of stress concentration.

WELDING FABRICATION & REPAIR 285

Stress concentrations at the corners of a square hole can be much larger and, small cracks and scratches that have sharp points have stress concentration factors between 100 and 1000. Many shipbuilders have learned about these factors the hard way. These concentrations are more than enough to cause crack growth and fatigue failure. Figure 7–29 shows a metal bar under tension and a shaft under torsion, respectively with stress concentrators and the same parts with modifications to reduce stress concentrations caused by these abrupt shape changes.

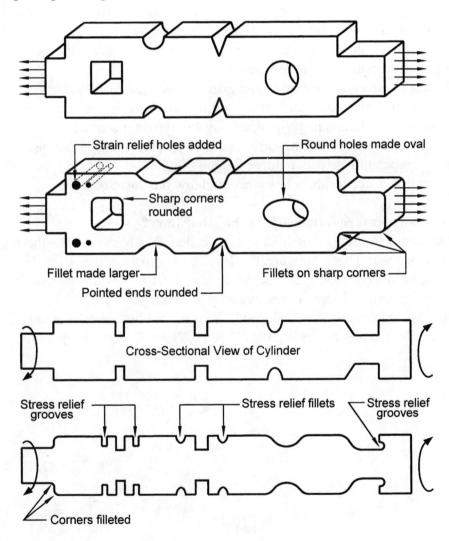

Figure 7–29. Typical stress concentrators and modifications to reduce stress concentrations in these parts. Upper two bars are in tension, lower two shafts are in torsion.

What are some of the important factors that influence fatigue life?
- Sharp corners and abrupt changes in shape create stress points.
- The larger the part, the greater the chances of fatigue cracks, because a greater volume of metal is under high stress and there is also a greater chance of residual stress increasing the stress the part sees. Residual stress is added to external stress in calculating fatigue stress.
- Type of stress loading tension, bending, or torsion affects results.
- Surface finish, particularly scratches, scribe marks, machining tool marks, and irregular points along a weld bead add points of stress concentration. Removing surface scratches by polishing may be helpful.
- Surface irregularities caused by rolling, forging, or stamping processes create stress points on the part surface.
- Surface treatments like plating and grinding increase surface stress.
- Fine-grained materials, particularly high-strength steels, are more sensitive to surface finish than larger-grained, lower-strength materials.
- Corrosion attacks the metal surface and causes stress concentrations.
- High temperature usually reduces fatigue life.
- If possible, place welds in low stress and low flex locations.

How can we measure fatigue life in the laboratory?
Plots of stress to number of load cycles, called *S-to-N curves* are the most common method. These measure the level of alternating stress on the part versus the number of load cycles before failure. Here is how this is done:
1. Multiple, identical samples are made up for testing.
2. Samples are fatigue tested. Depending on the application, the testing may be tensile, flexure, or torsional load cycles, Figure 7–30 through 7–32.

WELDING FABRICATION & REPAIR 287

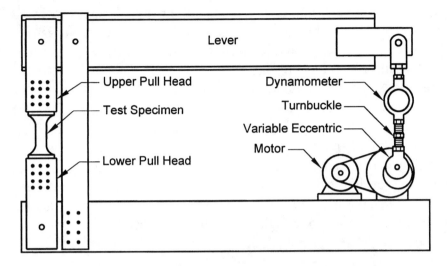

Figure 7–30. Testing machine for fatigue in tension.

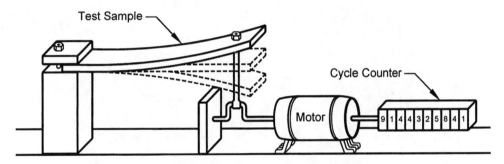

Figure 7–31. Testing machine for fatigue in flexing.

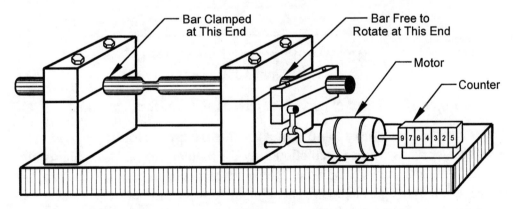

Figure 7–32. Testing machine for fatigue in torsion.

3. Some samples are subjected to load cycles at high stress, some at low stress, and some in between. The number of cycles-to-failure is recorded. The average of many samples run at the same stress level are needed to produce meaningful test results.
4. The S-to-N curve is plotted.
5. From the S-to-N curve, we can determine the level of stress for which the material has infinite life. Ferrous metals display a leveling off of the S-to-N curve and this point is called the fatigue limit, Figure 7–33 upper curve, which occurs at about 10^6 load cycles. On non-ferrous metals which do not have the leveling off of the S-to-N curve as ferrous metals do, the stress which will take the part to 5×10^8 cycles before failure is taken as the level for infinite life. See Figure 7–33 bottom curve.

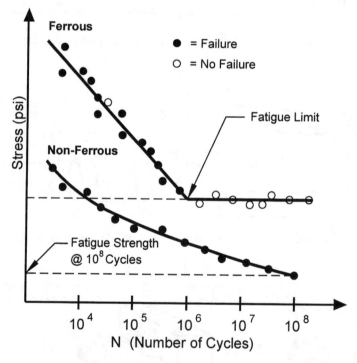

Figure 7–33. S-to-N curves for ferrous and non-ferrous metals.

What are the limitations to applying the S-to-N curve results?
- These curves give life prediction only for regular, repeated loading of the part and that may not be what the part really sees in service. One way to get around this problem is to specify a maximum number of service hours.
- S-to-N curves are specific to a particular stress ratio and surface finish, so it may be hard to apply the test results to a particular part.

- They provide only the *average life* and many samples have to be run to reach high confidence levels.
- Fatigue data obtained for a simple geometric shape may be hard or impossible to apply to a complex part. The only reliable solution may be to test the actual part in a fatigue machine.

If samples of the same steel alloy were prepared, and then tested in tension as a forging, stress parallel to rolling grain, and stress across the rolling grain, what would the results be on an S-to-N curve?
Figure 7–34 shows that the forging has the best performance and the stress-across-the-grain sample the poorest.

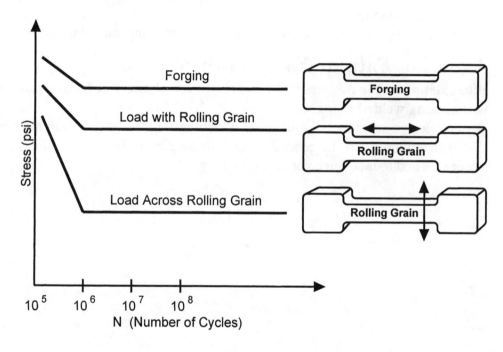

Figure 7–34. S-to-N curves for the same steel in tension.

What effect does adding a weld to a plate have on its fatigue strength?
The addition of a weld to the plate disrupts the smooth flow of stress through the plate, causing stress concentrations. These cause significant reduction in fatigue life as shown in Figure 7–35.

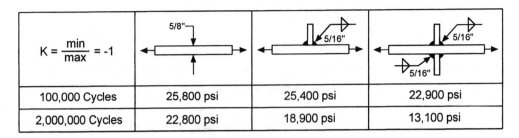

Figure 7–35. Effect of adding weld joint on plate fatigue strength.

What surface treatments improve fatigue life?
- Heat treatment processes like nitriding and carburizing produce higher strength materials.
- Shot peening undoes the effects of corrosion, grinding, and plating.

Avoiding Fatigue Failures through Design

What steps can the designer take to reduce the chances of fatigue failure in designing welded items?

We can design to reduce the chances of fatigue failure by the design steps shown in Figure 7–36. In general, we try to distribute flexing action over a larger area and reduce force concentrations.

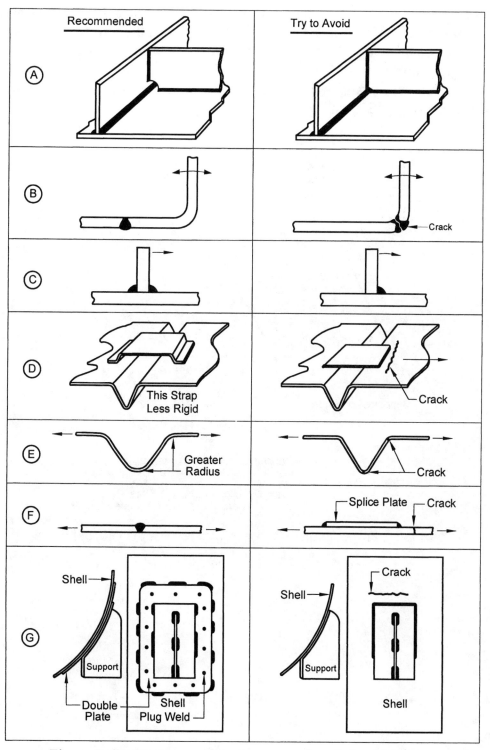

Figure 7–36. Design modifications to reduce fatigue failures.

In structures subject to fatigue or cyclical loading, how should the sheet or plate steel grain direction (the direction of hot rolling performed in the steel mill) be placed with respect to the direction of the cyclical load?

The grain direction should be placed *in line* with the direction of cyclical loading for longest fatigue life, Figure 7–37. An example of this application is the tank on a gasoline tank truck.

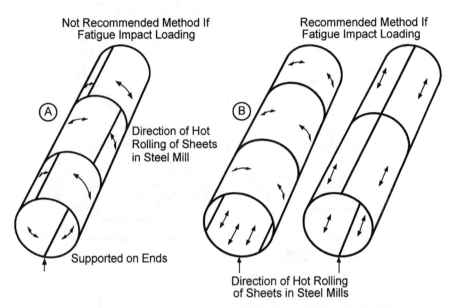

Figure 7–37. Proper grain orientation with respect to cyclical loading.

Chapter 8

Tools, Materials, Supplies & Information

*It isn't that they can't see the solution.
It is that they can't see the problem.*
—G.K Chesterton

Applied Bolting Technology Products, Inc.
Applied Bolting manufactures DTIs for structural steel bolts from ½- to 1½-inch diameter; they make DTIs for metric bolts also.
www.appliedbolting.com

Assembly Technologies International, Inc.
Manufactures American Beauty® resistance soldering equipment.
www.assemblytech.com

Bessey & Sohn GmbH & Co.
Offers a complete line of heavy duty welding clamps and accessories.
www.bessey.de

Chicago Metal Rolled Products Company
This company rolls angles, bars, beams, channels, tees, pipe, tube, and formed sections to a specified radius of a circle or segment of a circle. They stock many sizes of angle rings. Other items are available on a short delivery schedule. They offer high tolerances on bends and roll sheet and plate from 24 gauge to 5/16 inch (0.6 to 8 mm) into pipe, and cylinders to 12-foot (3.7 m) lengths.
www.cmrp.com

Copper Development Association, Inc.
This organization promotes the use of copper. It offers more than one hundred publications on copper and its applications, and has a series of linked web sites with excellent copper technical information.
www.copper.org

DE-STA-CO Industries
This company makes a complete line of quick-release clamps and clamping systems for fixtures and production equipment. They have an excellent catalog.
www.destaco.com

Di-Acro, Incorporated
This company manufactures a variety of sheet metal bending machines—manual, power, and CNC. They offer shears, rod parters, slip rolls, notchers, and punching systems. Their equipment is for light-to-medium sheet metal.
www.diacro.com

Enco Manufacturing Co.
Enco offers a full line of shop tools, machinery, and supplies both imported and domestic. They also offer wheels, axles, fenders, brakes, lights, and suspensions for the trailer builder.
www.use-enco.com

Haydon Bolts, Inc.
Haydon specializes in custom manufactured anchor bolts, tie rod assemblies, U-bolts, eye bolts, and J-bolts in diameters from 1/2 through 4 inches and up to 40 feet in length. Along with manufactured products, Haydon has one of the nation's largest stocks of tension control bolts (tension shear) and hex head bolts in A325 and A490. Haydon also sells and services wrenches for tension control and hex bolts.
www.haydonbolts.com

International Fastener Institute
This organization operates an excellent site on fastener technology, standardization, and related issues.
www.industrial-fasteners.org

Kant-Twist™ Clamp Manufacturing Co, Inc.
The Kant-Twist multipurpose lever clamps combine the best features of ordinary C-clamps and parallel clamps. They are available in many sizes and

Welding Fabrication & Repair

styles for every conceivable use. This company also offers industrial-grade hammers.
www.clampmfg.com

Mathey Dearman, Inc.
Supplies every type of hardware and tool for the pipeliner: pipe cutting and beveling machines, pipe alignment and reforming clamps, flange tools, pipefitter's tools, welding electrodes, flux ovens, and welding work benches.
www.mathey.com

Milwaukee Electric Tool Corporation
Offers a wide range of heavy-duty industrial-grade electric power tools for cutting, grinding, boring, polishing, hoisting, and drilling.
www.milwaukeeconnect.com

MK Products Inc.
MK manufactures orbital welding equipment wire feeders, spool guns, rotary tables, welding torches, and power supplies. Has an outstanding web site with a complete list of products, sales literature, owner's manuals, FAQs, and prices.
www.mkprod.com

Nucor Corporation Fastener Division
Supplies structural steel nuts and bolts, hex head cap screws, and hex nuts.
www.nucor-fastener.com

J. P. Nissen Company
Supplier of markers for metal and other materials.
www.nissenmarkers.com

L. S. Starrett Company
World-class manufacturer of precision measurement tools, layout tools, levels, hacksaw and band saw blades, and other metal cutting tools.
www.lsstarrett.com

Oxylance Corporation
Exothermic burning bars, underwater cutting lances, pipe holders, bare lance pipe, hoses, and high-pressure oxygen regulators.
www.oxylance.com

Reid Tool Supply Company
Industrial tools, shop supplies, comprehensive line of machine-building hardware such as knobs, springs, bearings, latches, handles, hand wheels, drawer slides, mounting feet, and fasteners. Also offer plastic, aluminum, brass, stainless steel, cast iron, and steel stock.
www.reidtool.com

Small Parts Inc.
Wide selection of small hand tools and a remarkable assortment of hard-to-find metal stock, plastic materials, tubing, fasteners, fittings, and mechanical components. Most items are for smaller-scale projects in medical, instrumentation, machinery, and model building.

www.smallparts.com

U.S. Industrial Tool & Supply Company
Riveting tools, metal cutting tools, light aircraft engine, and airframe maintenance tools.
www.ustool.com

Van Sant Enterprises, Inc.
This company offers a variety of sheet metal and metal working tools: metal brakes, metal benders, tubing benders, grinding and sanding machines, iron workers, tube coping machines and fixtures, and welding accessories.
www.vansantent.com

Vise-Grip® (American Tool Companies, Inc.)
American Tool offers nearly forty different designs of locking pliers for welding, plumbing, metalworking, and general shopwork.
www.americantool.com

Walhonde Tools, Inc.
This company provides patented boiler tube pulling and alignment tools, plus pipe alignment tools for pipe from 5- to 20-inch diameter for welding. They also offer their products for rent.
www.walhonde.com

Weldsale Company (Division of A.J. Cunningham Equipment Inc.)
Welding platens, tooling and clamping systems, cutting tables, cutting pyramids, positioners, and downdraft systems.
www.weldsale.com

Welding Fabrication & Repair

Williams Low Buck Tools
Economy source of tubing clamps, benders, notchers, metal forming tools, punches, and other metal working tools. Many uncommon tools.
www.lowbucktools.com

Appendices

A – Conversion Factors

	English to Metric	Metric to English
Capacity	$ft^3 \times 0.028316 = m^3$	$m^3 \times 35.315 = ft^3$
	$in^3 \times 0.01639$ = liters	liters $\times 61.02 = in^3$
	U.S. gallons $\times 3.785$ = liters	liters $\times 0.2642$ = U.S. gallons
	$in^3 \times 16.39 = cm^3$	$cm^3 \times 0.06102 = in^3$
Length	in $\times 25.40$ = mm	mm $\times 0.03937$ = in
Weight	lb $\times 0.453592$ = kg	kg $\times 2.20462$ = lb
Pressure	psi $\times 0.06895$ = bars	bars $\times 14.5$ = psi
	psi $\times 6.895$ = kPa	kPa $\times 0.145$ = psi
	psi $\times 0.006895$ = MPa	MPa $\times 145$ = psi
	psi $\times 0.0703 = kg/cm^2$	$kg/cm^2 \times 14.224$ = psi
Flow Rate	ft^3/min $\times 1699$ = liters/hr	liters/hr $\times 0.000589 = ft^3$/min

B – Important Elements in Metallurgy

Material	Density Specific Gravity	Weight (lb/ft³)	Melting Point °C	Melting Point °F
Bronze (90% Cu-10% Sn)	8.78	548	850-1000	1562-1832
Brass (90% Cu-10% Zn)	8.60	540	1020-1030	1868-1886
Brass (70% Cu-30% Zn)	8.44	527	900-940	1652-1724
Bronze (90% Cu-9% Al)	7.69	480	1040	1905
Bronze, Phosphor (90% Cu-10% Sn)	8.78	551	1000	1830
Bronze, Silicon (96% Cu-3% Si)	8.72	542	1025	1880
Iron, Cast	7.50	450	1260	2300
Iron, Wrought	7.80	485	1510	2750
Steel, high-carbon	7.85	490	1374	2500
Steel, low-alloy	7.85	490	1430	2600
Steel, low-carbon	7.84	490	1483	2700
Steel, medium-carbon	7.84	490	1430	2600

Glossary

A

acceptable weld: A weld that meets the applicable requirements.

acetone: A colorless, flammable, volatile liquid used as a paint remover and as a solvent for oils and other organic compounds. Used in acetylene cylinders to saturate the monolithic filler material to stabilize the acetylene.

acetylene feather: The intense white, feathery-edged portion adjacent to the cone of a carburizing oxyacetylene flame.

actual throat: The shortest distance between the weld root and the face of a fillet weld.

adhesion: A state of being stuck together. The joining together of parts that are normally separate.

AISI: The American Iron and Steel Institute.

alloy steel: A plain carbon steel to which another element, other than iron and carbon, has been added in a percentage large enough to alter its characteristics.

alloy: A substance with metallic properties, composed of two or more chemical elements of which at least one is a metal.

alloying element: Elements added in a large enough percentage to change the characteristics of the metal. Such elements may be chromium, manganese, nickel, tungsten, or vanadium; these elements are added to produce specific physical properties such as hardness, toughness, ductility, strength, resistance to corrosion, or resistance to wear.

allthread: Steel or stainless steel rod threaded from end to end. It is available in a wide range of diameters and thread specifications.

alternating current (AC): An electric current that reverses its direction periodically.

alternative fuels: Propane, methylacetylene propadiene (MPS), natural gas, or fuel gases, other than acetylene, used for welding or cutting.

aluminum: One of the chemical elements, a silvery, lightweight, easily worked metal that resists corrosion.

ampere: A unit of electrical current measuring the rate of flow of electrons through a circuit. One ampere is equivalent to the current produced by one volt applied across a resistance of one ohm.

annealing: A process of heating then cooling metal to acquire desired qualities such as ductility.

anode: The positive terminal of an electrical source.

anodizing: An electrochemical process that thickens and toughens the naturally occurring protective oxide on aluminum. The anodic coating is part of the metal, but has a porous structure which allows infusions or color or lubricants and provides a durable, attractive surface finish.

arc blow: The deflection of an arc from its normal path because of magnetic forces.

arc force: The axial force developed by an arc plasma.

arc gap: A nonstandard term used for the arc length.

arc gouging: Thermal gouging that uses an arc cutting process variation to form a bevel or groove.

arc plasma: A gas that has been heated by an arc to at least a partially ionized condition, enabling it to conduct electric current.

arc spraying: A thermal spraying process using an arc between two consumable electrodes of surfacing materials as a heat source and a compressed gas to atomize and propel the surfacing material to the substrate.

arc strike: A discontinuity resulting from an arc, consisting of any localized re-melted metal, heat-affected metal, or change in the surface profile of any metal object.

arc time: The time during which an arc is maintained in making an arc weld.

arc voltage: The voltage across the arc.

arc welding: Arc welding is a group of welding processes in which fusion is produced by heating with an electric arc or arcs with or without the application of pressure and with or without the use of filler metal.

as welded: The condition of weld metal, welded joint, and weldments after welding, but prior to any subsequent thermal or mechanical treatment.

ASME Pressure Vessel Code: Engineering design specifications, including welding procedures, used on pressure vessels to insure safety.

ASTM: The American Society for Testing and Materials.

austenite: One of the basic steel microstructures wherein carbon is dissolved in iron. Austenite forms at elevated temperatures.

autogenous weld: A fusion weld made without using a filler material.

AWS: The American Welding Society.

axis of a weld: A line through the length of a weld, perpendicular to and at the geometric center of its cross-section.

B

back bead: A weld bead resulting from a back weld pass. Back beads are made *after* the primary weld is completed.

back fire: The momentary recession of the flame into the welding tip, or cutting tip followed by immediate reappearance or complete extinction of the flame, accompanied by a loud popping report.

back weld: A weld made at the back of a single groove weld.

back gouging: The removal of weld metal and base metal from the weld root side of a welded joint to facilitate complete fusion and complete joint penetration upon subsequent welding from that side.

backhand welding: A welding technique in which the welding torch or gun is directed opposite to the progress of welding.

backing bead: A weld bead resulting from a backing pass. Backing beads are completed *before* welding the primary weld.

backing pass: A weld pass made to provide a backing for the primary weld.

backing ring: Metal ring used as backing in pipe welding to prevent burn-through. Backing rings may also aid in alignment of the two pipes being welded.

backing strip: Non-standard term used to describe a backing on the root side of the weld in the form of a strip.

backing: Material or device placed against the back side of a joint to support and retain molten weld-metal. The material may be partially fused or remain unfused during welding and may be either metal or nonmetal (metal strip, asbestos, carbon, copper, inert gas, ceramics).

back-step sequence: A longitudinal sequence in which weld passes are made in the direction opposite weld progression, usually used to control distortion.

bainite: A steel microstructure that is harder than pearlite, cementite, or ferrite, and more ductile than martensite.

base material: The material that is welded, brazed, soldered, or cut.

base metal: The metal or alloy that is welded, brazed, soldered, or cut.

bead weld: A term used for surfacing welds.

bend allowance: Additional length added to a part to provide the extra material needed to go around bend radii. Failure to add bend allowance material prevents making satisfactory bends.

bevel angle: The angle between the bevel of a joint member and a plane perpendicular to the surface of the member.

bevel: An edge preparation, the angular edge shape.

body-centered cubic (BCC): One of the common types of unit cells described as a cube with an atom at each of the eight corners and a single atom at the center of the cell. This arrangement is typical of the ferritic form of iron. Among the common BCC metals are iron, carbon steel, chromium, molybdenum, and tungsten.

boxing: The continuation of a fillet weld around a corner of a member as an extension of the principle weld.

braze metal: The filler metal used to make the joint of a brazed socket having a liquidus above 850°F (450°C).

braze welding: A welding process that uses a filler metal with a liquidus above 840°F (450°C) and below solidus of the base metal. The base metal is not melted. Unlike brazing, in braze welding the filler metal is not distributed in the joint by capillary action.

braze: A weld produced by heating an assembly to the brazing temperature using filler metals having a liquidus above 840°F (450°C) and below the solidus of the base metal. The filler metal is distributed between the closely fitted faying surfaces of the joint by capillary action.

brazing filler metal: The metal or alloy used as a filler metal in brazing, which has liquidus above 850°F (450°C) and below the solidus of the base metal.

brazing: A group of welding processes that produces coalescence of materials by heating them to the brazing temperature in the presence of a filler metal having a liquidus above 850°F (450°C) and below the solidus of the base metal. The filler metal is distributed between the closely fitted faying surfaces of the joint by capillary action.

Brinell hardness test: A common testing method using a ball penetrator. The diameter of the indentation is converted to units of Brinell hardness number (BHN).

buckling: Bending or warping caused by the heat of welding.

butt joint: A joint between two members aligned approximately in the same plane.

buttering: A surfacing variation that deposits surfacing metal on one or more surfaces to provide metallurgically compatible weld metal for the subsequent completion of the weld.

butting member: A joint member that is prevented by the other member from movement in one direction perpendicular to its thickness dimension.

C

capacitor: A device consisting of two or more conducting plates separated from one another by an insulating material and used for storing an electrical charge.

capillary action: The force by which liquid in contact with a solid is distributed between closely fitted faying surfaces of the joint to be brazed or soldered.

carbon steel: A steel containing various percentages of carbon. Low-carbon steel contains a maximum of 0.15% carbon; mild steel contains 0.15% to 0.35% carbon; medium-carbon steel contains 0.35% to 0.60% carbon; high-carbon steel contains from 0.60% to 1.0% carbon.

carbon: A nonmetallic chemical element that occurs in many inorganic and all organic compounds. Carbon is found in diamond and graphite, and is a constituent of coal, petroleum, asphalt, limestone, and other carbonates. In combination, it occurs as carbon dioxide and as a constituent of all living things. Adjustment of the amount of carbon in iron produces steel.

carburizing flame: A reducing oxygen-fuel gas flame in which there is an excess of fuel gas, resulting in a carbon-rich zone extending around and beyond the inner cone of the flame.

cast iron: A family of alloys, containing more than 2% carbon and between 1% and 3%

silicon. Cast irons are not malleable when solid, and most have low ductility and poor resistance to impact loading. There are four basic types of cast iron gray, white, ductile, and malleable.

cathode: The negative terminal of a power supply; the electrode when using direct current electrode negative (DCEN).

caulking: Plastic deformation of weld and adjacent base metal surfaces by mechanical means to seal or obscure discontinuities.

cementite: A very hard form of low-temperature steel that contains more than 0.8% carbon. Cementite occurs in steel that has not been previously heat treated or in steel that has been cooled slowly after being transformed into austenite.

chain intermittent weld: An intermittent weld on both sides of a joint where the weld increments on one side are approximately opposite those on the other side.

Charpy V-notch test: An impact test used to determine the notch toughness of materials.

chill plate: A piece of metal placed behind material being welded to correct overheating.

chill ring: A non-standard term for a backing ring.

chromium: A lustrous, hard, brittle, steel-gray metallic element used to harden steel alloys, in production of stainless steel, and as a corrosion resistant plating.

cladding: A surfacing variation that deposits or applies surfacing material usually to improve corrosion or heat resistance.

coalescence: The growing together or growth into one body of the materials being welded.

coefficient of thermal expansion: The increase in length per unit length for each degree a metal is heated.

cohesion: Cohesion is the result of a perfect fusion and penetration when the molecules of the parent material and the added filler materials thoroughly integrate as in a weld.

cold crack: A crack that develops after solidification is complete.

cold soldered joint: A joint with incomplete coalescence caused by insufficient application of heat to the base metal during soldering.

cold work: Cold working refers to forming, bending, or hammering a metal well below the melting point. Cold working of metals causes hardening, making them stronger but less ductile.

complete fusion: Fusion over the entire fusion faces and between all adjoining weld beads.

complete joint penetration: A root condition in a groove weld in which weld metal extends through the joint thickness.

composite electrode: A generic term for multi-component filler metal electrodes in various physical forms such as stranded wires, tubes, or covered wire.

composite: A material consisting of two or more discrete materials with each material retaining its physical identity.

concavity: The maximum distance from the face of a concave fillet weld perpendicular to a line joining the weld toes. A concave fillet weld will have a face that is contoured below a straight line between the two toes of a fillet weld.

conductor: A device, usually a wire, used to connect or join one circuit or terminal to another.

cone: The conical part of an oxygen-fuel gas flame adjacent to the tip orifice.

constant-current (CC) power source: An arc welding power source with a volt-ampere relationship yielding a small welding current change from a large arc voltage change.

constant-voltage (CV) power source: An arc welding power source with a volt-ampere relationship yielding a large welding current change from a small arc voltage change.

constricted arc: A plasma arc column that is shaped by the constricting orifice in the nozzle of the plasma arc torch or plasma spraying gun.

consumable backing ring: Backing ring that is completely fused into the weld joint metal during welding.

consumable electrode: An electrode that provides filler metal, therefore is consumed in the arc welding process.

consumable insert: Filler metal that is placed at the joint root before welding, and is intended to be completely fused into the joint root to become part of the completed weld.

contact resistance: Resistance to the flow of electric current between two work-pieces or an electrode and the work-piece.

contact tube setback: The distance from the contact tube to the end of the gas nozzle. This term is used in gas metal arc and gas shielded flux cored arc welding.

contact tube: A device that transfers current to a continuous electrode.

continuous welding (CW): ASW nomenclature for an electrical welding process used to close the seam on steel pipe formed from flat strip.

convexity: The maximum distance from the face of a convex fillet weld perpendicular to a line joining the toes.

corner joint: A joint between two members located approximately at right angles to each other in the form of an *L*.

corrosive flux: A flux with a residue that chemically attacks the base metal. It may be composed of inorganic salts and acids, organic salts and acids, or activated rosin.

cosmetic pass: A weld pass made primarily to enhance appearance.

cover pass: The final bead(s) of a multi-pass weld.

cover plate: A removable pane of colorless glass, plastics coated glass, or plastics that covers the filter plate and protects it from weld spatter, pitting, or scratching.

covered electrode: A composite filler metal electrode consisting of a core of a bare electrode or metal cored electrode to which a covering sufficient to provide a slag layer on the weld metal has been applied. The covering may contain materials providing such functions as shielding from atmosphere, deoxidation, and arc stabilization, and can serve as a source of metallic additions to the weld.

crack: A fracture-type discontinuity characterized by a sharp tip and high ratio of length and width to opening displacement.

cracking a valve: Rapidly opening and closing a valve to clear the orifice of unwanted foreign material.

crater crack: Radial cracks formed in a weld crater as the weld pool solidifies and shrinks.

crater: A depression in the weld face at the termination of a weld bead.

crystalline structure: The orderly arrangement of atoms in a solid in a specific geometric pattern. Sometimes called a lattice structure.

cutting attachment: A device for converting an oxygen-fuel gas welding torch into an oxygen-fuel cutting torch.

cutting head: The part of a cutting attachment to which the cutting torch or tip may be attached.

cutting pyramid: A pyramidal shaped, cast-iron spacer used to hold plate goods 2 to 4 inches *above* the cutting table. Cutting pyramids prevent torch damage to the cutting table, and avoid the flame deflection that occurs with supports of other shapes.

cutting tip: An attachment to an oxygen cutting torch from which the gases exit.

cycle: The duration of alternating current represented by the current increase from an initial value to a maximum in one direction then to a maximum in the reverse direction and its return to the original initial value.

cylinder manifold: A multiple header for interconnection of gas sources with distribution points.

D

defect: A discontinuity or discontinuities that by nature or accumulated effect render a part or product unable to meet minimum applicable acceptance standards or specifications. The term designates rejection.

deposited metal: Filler metal that has been added during welding, brazing, or soldering.

deposition rate: The weight of filler material deposited in a unit of time.

depth of bevel: The perpendicular distance from the base metal surface to the root edge or the beginning of the root face.

depth of fusion: The distance that fusion extends into the base metal or previous bead from the surface melted during welding.

diameter nominal (DN): Each nominal pipe size in inch-pound units has its metric equivalent called the *diameter nominal*. Both "25 mm" and "2 inch" are names the trade gives to pipes of 2-3/8 inch OD. Fittings, flanges, couplings, valves, as well as other piping components, are interchangeable in the two measurement systems thanks to the work of the International Standards Organization.

diode: Diodes are check valves for electricity. They will pass current in only one direction, from plus to minus, and are used to convert AC to DC.

dip brazing (DB): A brazing process that uses heat from a molten salt or metal bath. When a molten salt is used, the bath may act as a flux. When a molten metal is used, the bath provides the filler metal.

direct current electrode negative (DCEN): The arrangement of direct current arc welding cables in which the electrode is the negative pole and the workpiece is the positive pole of the welding arc.

direct current electrode positive (DCEP): The arrangement of direct current arc welding cables in which the electrode is the positive pole and the workpiece is the negative pole of the welding arc.

discontinuity: An interruption of the typical structure of a material, such as a lack of homogeneity in its mechanical, metallurgical, or physical characteristics. A discontinuity is not necessarily a defect.

distortion: Non-uniform expansion and contraction of metal caused by heating and cooling during the welding process.

dog: A steel tool used to hold work in position during welding or bending operations. Usually the dog is held in position by the wedging action of the forces on it.

double submerged arc welding (DSAW): ASW nomenclature for a fully automatic consumable wire electrode welding process used on thick plate. DSAW is performed under a layer of granular flux and two electrode wires are used simultaneously.

downhill: Welding in a downward progression.

drag angle: The travel angle when the electrode is pointing in a direction opposite to the progression of welding. This angle can also be used to partially define the positions of guns, torches, and rods.

drag: During thermal cutting, the offset distance between the actual and straight line exit points of the gas stream or cutting beam measured on the exit surface of the base metal.

ductility: The tendency to stretch or deform appreciably before fracturing.

duty cycle: The percentage of time during an arbitrary test period that a power source or its accessories can be operated at rated output without overheating. Most welding machines are rated in intervals of ten minutes meaning that a duty cycle of 50% means the machine can be operated at a given amperage setting for five continuous minutes without damage to the equipment. 60% would give six minutes; 70% would give seven minutes.

E

edge joint: A joint between the edges of two or more parallel or nearly parallel members.

edge preparation: The preparation of the edges of the joint members, by cutting, cleaning, plating, or other means.

effective throat: The minimum distance, minus any convexity, between the weld root and the face of a fillet weld.

elastic limit: The greatest stress the material can withstand without permanent elongation when all load has been removed from the specimen. Any strains developed up to the elastic limit are both small and reversible.

electrical resistance soldering: A soldering process which uses heat generated by the passage of an electric current through the work or surrounding carbon electrodes to replace an open flame. It can be used on copper tubing and a variety of other applications.

electrode angle: The angle of the electrode in relationship to the surface of the material being welded; the electrode's perpendicular angle to the metals' surface leaning toward the direction of travel.

electrode classification: A means of identifying electrodes by their usability, flux coverings, and chemical make up. The American Welding Society has published a series of specifications for consumables used in welding processes.

electrode extension: In gas metal arc welding, flux cored arc welding, electrogas welding, and submerged arc welding, it is the length of electrode extending beyond the end of the contact tube. In gas tungsten arc welding and plasma arc welding, it is the length of the tungsten extending beyond the end of the collet.

electrode holder: A device used for mechanically holding and conducting current to an electrode during welding or cutting.

electrode lead: The electrical conductor between the source of arc welding current and the electrode holder.

electrode: A component of the electrical welding circuit that terminates at the arc, molten conductive slag, or base metal.

elongation: The amount of permanent extension in the vicinity of a fracture in a tension test; usually expressed in a percentage of original gauge length.

eutectic composition: The composition of an alloy system that has two descending liquidus curves; the lowest possible melting point for that mixture of metals.

F

face bend test: A test in which the weld face is on the convex surface of a specified bend radius.

face reinforcement: Weld reinforcement on the side of the joint from which welding was done.

face-centered cubic (FCC): One of the common types of unit cells in which atoms are located on each corner and the center of each face of a cube. Among the common FCC metals are aluminum, copper, nickel, and austenitic stainless steel. This arrangement is typical of the austenitic form of iron.

fatigue failure: The sudden and complete breakage of a part as a result of the repeated application of a load. Fatigue failure is progressive and may not occur until after millions of load cycles.

fatigue strength: Ability of a material to withstand repeated loading.

faying surface: The mating surface of a member that is in contact with or in close proximity to another member to which it is to be joined.

ferrite: A form of low-temperature steel that contains a very small percentage of carbon. Ferrite occurs in steel that has not been previously heat treated or in steel that has been cooled slowly after being transformed to austenite.

filler material: The material, metal, or alloy to be added in making a welded, brazed, or soldered joint.

filler metal: The metal also known as brazing filler metal, consumable insert, diffusion aid, filler material, solder, welding electrode, welding filler metal, welding rod, and welding wire.

filler pass(es): The welding passes following the root and hot pass. These passes deposit the majority of the filler metal. The last filler pass lies under the cover pass.

fillet weld break test: A test in which the specimen is loaded so that the weld root is in tension.

fillet weld leg: The distance from the joint root to the toe of the fillet weld.

fillet weld size: For equal-leg fillet welds, the leg lengths of the largest isosceles right triangle that can be inscribed within the fillet weld cross-section. For unequal leg fillet welds, the leg lengths of the right triangle can be inscribed within the fillet weld cross-section.

fillet weld: A weld of approximately triangular cross-section joining two surfaces approximately at right angles to each other in a lap joint, T-joint, or corner joint.

filter plate shade: Refers to the lens darkness number, which indicates the darkness of the lens.

filter plate: An optical material that protects the eyes against excessive ultraviolet, infrared, and visible radiation. Also called filter glass or filter lens.

Five F (5F): A welding test position designation for a circumferential fillet weld applied to a joint in pipe, with its axis approximately horizontal, in which the weld is made in the horizontal, vertical, and overhead welding positions. The pipe remains fixed until the welding of the joint is complete.

Five G (5G): A welding test position designation for a circumferential groove weld applied to a joint in a pipe with its axis horizontal, in which the weld is made in the flat, vertical, and overhead welding positions. The pipe remains fixed until the welding of the joint is complete.

fixture: A device designed to hold and maintain parts in proper relation to each other.

fixture: A device or structure which securely holds the workpiece in position and alignment while cutting or welding operations are performed.

flame bending: A process using oxyfuel flame heat to expand and upset the work piece forming a bend.

flame propagation rate: The speed at which flame travels through a mixture of gases.

flame shrinking: A process using oxyfuel flame heat to expand and upset the workpiece metal to remove wrinkles from the interior of welded panels.

flame straightening: A process using oxyfuel flame heat to expand and upset the workpiece metal differentially to reform beams or columns into their original shape. Many heating and cooling cycles are usually needed.

flared tubing connection: A means of connecting tubing to a fitting by forcing a flared tubing end against a tapered surface making a liquid-tight seal.

flare-V-groove weld: A weld in a groove formed by two members with curved surfaces.

flashback arrester: A device to limit damage from a flashback by preventing propagation of the flame from beyond the location of the arrester.

flashback: A recession of the flame into or back of the mixing chamber of the oxygen fuel gas torch or flame spraying gun.

flat welding position: The welding position used to weld from the upper side of the joint at a point where the weld axis is approximately horizontal, and the weld face lies in an approximately horizontal plane.

flaw: An undesirable blemish or discontinuity in a weld such as a crack or porosity.

flux cored electrode: A composite tubular filler metal electrode consisting of a metal sheath and a core of various powdered materials producing an extensive slag cover on the face of a weld bead. External shielding may be required.

flux cutting (FOC): An oxygen cutting process that uses heat from an oxyfuel gas flame with a flux in the flame to aid cutting.

flux: A material used to hinder or prevent the formation of oxides and other undesirable substances in molten metal and on solid metal surfaces, and to dissolve or otherwise facilitate the removal of such substances.

forehand welding: A welding technique in which the welding torch or gun is directed toward the progress of welding.

fuel gas: A gas such as acetylene, natural gas, hydrogen, propane, stabilized methylacetylene propadiene, and other fuels normally used with oxygen in one of the oxyfuel processes and for heating.

furnace brazing (FB): A brazing process in which the work-pieces are placed in a furnace and heated to the brazing temperature.

furnace soldering (FS): A soldering process in which the work-pieces are placed in a furnace and heated to the soldering temperature.

fusible plug: A metal alloy plug that closes the discharge channel of a gas cylinder and is designed to melt at a predetermined temperature permitting the escape of gas.

fusion face: A surface of the base metal that will be melted during welding.

fusion welding: Any welding process that uses fusion of the base metal to make the weld.

fusion zone: The area of base metal as determined on the cross-section of a weld.

fusion: The joining of base material, with or without filler material, by melting them together.

G

gas cylinder: A portable container used for transportation and storage of compressed gas.

gas nozzle: A device at the exit end of the torch or gun that directs shielding gas.

gas regulator: A device for controlling the delivery of gas at some substantially constant pressure.

globular transfer: In arc welding, the transfer of molten metal in large drops from a consumable electrode across the arc.

GMAW: The welding process Gas Metal Arc Welding; non-standard terms for this process are MIG (metal inert gas), MAG (metal active gas), wire feed, hard wire welding.

goggles: Protective glasses equipped with filter plates set in a frame that fits snugly against the face and used primarily with oxygen fuel gas welding processes.

groove angle: The total included angle of the groove between workpieces.

groove face: The surface of a joint member included in the side of the groove from root to toe.

groove radius: The radius used to form the shape of a J- or U-groove weld.

groove weld size: The joint penetration of a groove weld. Also groove throat or effective throat.

groove weld: A weld made in a groove between the workpieces. See welding symbols.

ground connection: An electrical connection of the welding machine frame to the earth for safety.

GTAW: The Gas Tungsten Arc Welding process; non-standard terms are Heliarc™, and TIG (tungsten inert gas).

H

hardfacing: A surfacing variation in which hard material is deposited to reduce wear.

HASTELLOY®: An expensive alloy of Ni, Mo, Cr, W, V, and Fe used where high corrosion and chemical resistance are required.

heat-affected zone (HAZ): The portion of the base metal whose mechanical properties or microstructure have been altered by the heat of welding, brazing, soldering, or thermal cutting.

hexagonal close packed (HCP): A unit cell in which two hexagons (six-sided shapes) form the top and bottom of the prism. An atom is located at the center and at each point of the hexagon. Three atoms, one at each point of a triangle, are located between the top and bottom hexagons. Among the common HCP metals are zinc, cadmium, and magnesium.

high carbon steel: See carbon steel.

horizontal welding position: In a fillet weld, the welding position in which the weld is on the upper side of an approximately horizontal surface and against an approximately vertical surface. In a groove weld, the welding position in which the weld face lies in an approximately vertical plane and the weld axis at the point of welding is approximately horizontal.

hot pass: The welding pass following the root pass that is performed at a higher than normal amperage and travel speed. Its objectives are to insure complete fusion of the root pass, to bring any entrapped slag to the surface for removal, and to produce a convex weld bead which is easy to clean.

hydrogen: The lightest chemical element, colorless, odorless, and tasteless. It is found in combination with other elements in most organic compounds and many inorganic compounds. Hydrogen combines readily with oxygen in the presence of heat, and forms water.

I

impact strength: The ability of a material to resist shock, dependent on both strength and ductility of the material.

inclusion: Entrapped foreign solid material, such as slag, flux, tungsten, or oxide.

incomplete fusion: A weld discontinuity in which fusion did not occur between weld metal and fusion faces or adjoining weld beads.

incomplete joint penetration: A joint root condition in a groove weld in which weld metal does not extend through the joint thickness.

inert gas: A gas that normally does not combine chemically with other elements or compounds.

infrared radiation: Electromagnetic energy with wave lengths 770 to 12,000 nanometers.

injector torch: An injector-type torch is used to increase the effective use of fuel gases supplied at pressures of 2 psi (14 kPa), or lower. The oxygen is supplied at pressures ranging from 10 to 40 psi (70 to 275 kPa), the pressure increasing to match the tip size. The relatively high velocity of the oxygen flow is used to aspirate or draw in more fuel gas than would normally flow at the low supply pressures of the fuel gases.

intermittent weld: A weld in which the continuity is broken by recurring unwelded spaces.

interpass temperature: In a multipass weld, the temperature of the weld area between weld passes.

inverter power supply: A welding power supply with solid-state electrical components that change the incoming 60 Hz power to a higher frequency. Changing the frequency results in greatly reducing the size and weight of the transformer. Inverter power supplies can be used with all arc welding processes.

iron carbide: A binary compound of carbon and iron; it becomes the strengthening constituent in steel.

iron carbon phase diagram: A graphical means of identifying different structures of steel and percentages of carbon occurring in steel at various temperatures.

isothermal transformation diagram: A graph that identifies different austenitic transformation products that occur over a period of cooling time at isothermal conditions. Also referred to as I-T diagram and T-T-T diagram meaning time-temperature-transformation diagram.

Izod test: A test performed on a specimen of metallic material to evaluate resistance to failure at a discontinuity and evaluate the resistance of a comparatively brittle material during extension of a crack. The test is performed using a small bar of round or square cross-section held as a cantilevered beam in a gripping anvil of a pendulum machine. The specimen is broken by a single overload impact of the swinging pendulum, and the energy absorbed in breaking the specimen is recorded by a stop pointer moved by the pendulum.

J

J-groove weld: A type of groove weld where one side of the joint forms a *J*.

jig: A device that holds the workpiece in place and guides the cutting tool during the cutting operation.

joint clearance: The distance between the faying surfaces of a joint in brazing or soldering.

joint design: The shape, dimensions, and configuration of the joint.

joint efficiency: The ratio of strength of a joint to the strength of the base metal expressed in percent.

joint filler: A metal plate inserted between the splice member and thinner joint member to accommodate joint members of dissimilar thickness in a spliced butt joint.

joint geometry: The shape and dimensions of a joint in cross-section prior to welding.

joint penetration: The distance the weld metal extends from the weld face into a joint, exclusive of weld reinforcement.

joint root: That portion of a joint to be welded where the members approach closest to each other. In cross-section, the joint root may be either a point, a line, or an area.

joint spacer: A metal part, such as a strip, bar, or ring, inserted in the joint root to serve as a backing and to maintain the root opening during welding.

joint type: A weld joint classification based on five basic joint configurations such as a butt joint, corner joint, edge joint, lap joint, and T-joint.

joint: The junction of members or the edges of members that are to be joined or have been joined.

K

kerf: The width of a cut produced during a cutting process.

keyhole welding: A technique in which a concentrated heat source penetrates partially or completely through a workpiece, forming a hole (or keyhole) at the leading edge of the weld pool. As the heat source progresses, the molten metal fills in behind the hole to form the weld bead.

killed steel: A molten steel that has been held in a ladle, furnace, or crucible until no more gas is evolved and the metal is perfectly quiet.

L

lamellar tear: A subsurface terrace and step-like crack in the base metal with a basic orientation parallel to the wrought surface. Such items are caused by tensile stresses in the through-thickness direction of the base metals when they have been weakened by the presence of small dispersed, planar shaped, nonmetallic inclusions parallel to the metal surface.

lamination: A type of discontinuity with separation or weakness generally aligned parallel to the worked surface of a metal.

lap joint: A joint between two overlapping members in parallel planes.

laser beam cutting (LBC): A thermal cutting process that severs metal by locally melting or vaporizing with the heat from a laser beam.

laser beam welding: A welding process that produces coalescence with the heat from a laser beam impinging on the joint. The process is used without a shielding gas and without the application of pressure.

laser: A device that produces a concentrated, coherent light beam by stimulated electronic or molecular transitions to lower energy levels. Laser is an acronym for *L*ight *A*mplification by *S*timulated *E*mission of *R*adiation.

lens shade: See filter plate shade.

line heating: The process of applying oxyfuel flame heat and water cooling to metal plate to form it into concave or convex shapes. It is often used in shipbuilding.

linear discontinuity: A discontinuity with a length that is substantially greater than its width.

liquidus: The lowest temperature at which a metal or an alloy is completely liquid.

longitudinal crack: A crack with its major axis orientation approximately parallel to the weld axis.

M

macroetch test: A test in which a specimen is prepared with a fine finish, etched, and examined under low magnification.

mandrel bending: A bending process that uses a die inside the pipe or tube during the bending operation to prevent wall deformation.

manganese: A gray-white nonmagnetic metallic element resembling iron, except it is harder and more brittle. Manganese can be alloyed with iron, copper, and nickel, for commercial alloys. In steel it increases hardness, strength, wear resistance, and other properties. Manganese is also added to magnesium-aluminum alloys to improve corrosion resistance.

manual welding: Welding with the torch, gun, or electrode holder held and manipulated by hand. Accessory equipment, such as part motion devices and manually controlled filler material feeders may be used.

MAPP® gas: A trade name for a fuel gas methacetylene-propadiene.

martensite: A very hard, brittle microstructure of steel produced when steel is rapidly quenched after being transformed into austenite.

mechanized welding: Welding with equipment that requires manual adjustment of the equipment controls in response to visual observation of the welding with the torch, gun, or electrode holder held by a mechanical device.

medium steel: Refer to carbon steel.

melt-through: Visible root reinforcement produced in a joint welded from one side.

metal electrode: A filler or non-filler metal electrode used in arc welding or cutting, which consists of a metal wire or rod that has been manufactured by any method and that is either bare or covered with a suitable covering or coating.

metal: A class of chemical elements that are good conductors of heat and electricity, usually malleable, ductile, lustrous, and more dense than other elemental substances.

metal-cored electrode: A composite tubular filler metal electrode consisting of a metal sheath and a core of various powdered materials.

metallic bond: The principal atomic bond that holds metals together.

metallurgy: The science explaining the properties, behavior, and internal structure of metals.

methylacetylene propadiene: A family of alternative fuel gases that are mixtures of two or more gases (propane, butane, butadiene, methylacetylene, and propadiene). Methylacetylene propadiene is used for oxyfuel cutting, heating, brazing, and soldering.

microetch test: A test in which the specimen is prepared with a polished finish, etched, and examined under high magnification.

microstructure: A term use to describe the structure of metals. Visual examination of etched metal surfaces and fractures reveal some configurations in etched patterns that relate to structure, but magnification of minute details yields considerably more information. Microstructures are examined with low-power magnifying glass, optical microscope, or electron microscope.

mild steel: Refer to carbon steel.

mixing chamber: That part of a welding or cutting torch in which a fuel gas and oxygen are mixed.

modulus of elasticity: The ratio of stress to strain in material; also referred to as Young's modulus.

molybdenum: A hard, silver-white metal, a significant alloying element in producing engineering steels, corrosion resistant steels, tool steels, and cast irons. Small amounts alloyed in steel promote uniform hardness and strength.

moment of inertia: A number that provides a measure of beam stiffness based on its size and shape. Moment of inertia has units of inches^4 or mm^4.

monel: A high tensile strength Ni-Cu alloy that exhibits high fatigue resistance in salt water, corrosive atmospheres, and various acid and alkaline solutions. It is non-magnetic and spark-resistant.

multipass welding: A weld requiring more than one pass to ensure complete and satisfactory joining of the metal pieces.

multiple welding position: An orientation for a non-rotated circumferential joint requiring welding in more than one welding position, as in welding a pipe or tube in a fixed position. (5F, 5G position is pipe on a horizontal plane and not moved or turned during welding; 6F, 6G position is pipe at 45° off the horizontal plane and not moved or turned during welding).

N

NEMA: The National Electrical Manufacturers' Association.

neutral flame: An oxyfuel gas flame that has characteristics neither oxidizing nor reducing.

nitrogen: A gaseous element that occurs freely in nature and constitutes about 78% of earth's atmosphere. Colorless, odorless, and relatively inert, although it combines directly with magnesium, lithium, and calcium when heated with them. Produced either by liquefaction and fractional distillation of air, or by heating a water solution of ammonium nitrate.

nominal pipe size (NPS): The size of pipe is identified by its *nominal pipe size.* For pipes between 1/8- and 12-inches nominal size, the outside diameter (OD) was originally selected so that the inside diameter was equal to the nominal size for pipes of standard wall thickness of the times. This is no longer true with the changes in metals and manufacturing processes, but the nominal size and standard OD continue in use as a trade standard.

non-consumable electrode: An electrode that does not provide filler metal, as used in the GTAW process.

non-corrosive flux: A soldering flux that in either its original or residual form does not chemically attack the base metal. It usually is composed of rosin-based materials.

non-destructive examination: The act of determining the suitability of some material or component for its intended purpose using techniques that do not affect its serviceability.

normalizing: The process of heating a metal above a critical temperature and allowing it to cool slowly under room temperature conditions to obtain a softer and less distorted material.

O

ohm: A unit of electrical resistance. An ohm is equal to resistance of a circuit in which a potential difference of one volt produces a current of one ampere.

open root joint: An unwelded joint without backing or consumable insert.

open-circuit voltage: The voltage between the output terminals of the power source when no current is flowing to the torch or gun.

orbital welding: A fully-automatic GTAW process for welding pipe and tube. It is most often used on stainless steel tubing.

overlap: The protrusion of weld metal beyond the weld toe or weld root.

oxidizing flame: An oxyfuel flame in which there is an excess of oxygen, resulting in an oxygen-rich zone extending around and beyond the cone.

oxygen lance: A length of pipe used to convey oxygen to the point of cutting in oxygen lance cutting.

oxygen: A colorless, odorless, tasteless, gaseous chemical element, the most abundant of all elements. Oxygen occurs free in the atmosphere, forming 1/5 of its volume, and in combination in water, sandstone, limestone, etc.; it is very active being able to combine with nearly all other elements and is essential to life.

oxyhydrogen cutting (OFC-H): An alternative fuel gas cutting process that uses hydrogen as the fuel source.

oxyhydrogen welding (OHW): An alternative fuel gas welding process that uses hydrogen as the fuel source.

oxynatural gas cutting (OFC-N): An alternative fuel gas cutting process that uses natural gas as the fuel source.

oxypropane cutting (OFC-P): An alternative fuel gas cutting process that uses propane gas as the fuel source.

P

parent metal: A non-standard term referring to the base metal.

partial joint penetration weld: A joint root condition in a groove weld in which incomplete joint penetration exists.

pass: A single progression of welding along a joint, resulting in a weld bead or layer.

pattern: Wood, metal, paper, or plastic sheet material that replicates the shape of a part. Patterns are used to transfer this shape to the work and may contain other information such as hole location, alignment marks, and bend lines.

pearlite: A mixture of ferrite and cementite that contains approximately 0.8% carbon. Pearlite occurs in low-temperature steel that has not been previously heat treated or in steel that has been cooled slowly after being transformed into austenite.

peening: The mechanical working of metals using impact blows.

penetration: A non-standard term used in describing depth of fusion, joint penetration, or root penetration.

phase diagram: A graph that identifies alloy phases occurring at various temperatures and percentages of alloying elements. Also referred to as an equilibrium diagram.

phase transitions: When metals or metal alloys go from solid to liquid or the reverse, this is a phase transition. Iron phase transitions are: at room temperature to 1,670°F (910°C) iron is body-center cubic, 1670°F (910°C) to 2535°F (1388°C) iron is face-center cubic, and 2535°F (1390°C) the melting point of iron to 2800°F (1538°C) iron is again body-center cubic. These changes are also called allotropic transformations.

phosphoric acid: The acid, H_3PO_4, widely used in industrial metal cleaning.

phosphorous: A highly reactive, toxic, nonmetallic element used in steel, glass, and pyrotechnics. It is almost always found in combination with other elements such as minerals or metal ores. Found in steel and cast iron as an impurity. In steel it is reduced to 0.05% or less otherwise phosphorous causes embrittlement and loss of toughness, however small amounts in low-carbon steel produce a slight increase in strength and corrosion resistance.

pilot arc: A low-current arc between the electrode and the constricting nozzle of the plasma arc torch to ionize the gas and facilitate the start of the welding arc.

pin fixture: A tool for bending wire, bar, or rod into a curve or series of curves.

pipe alignment tool: Metal fixture specifically designed to hold two pipes in place during welding. Some alignment tools can apply force against pipe walls to bring them back into round.

pipe flange: The enlarged structure on the end of a pipe that is designed for attaching one pipe to another or to a vessel.

pipe jacks: A pipe fitting tool used to support pipe or tubing at a convenient working height while performing welding and related operations.

plasma arc cutting (PAC): An arc cutting process that uses a constricted arc and removes the molten metal with a high-velocity jet of ionized gas issuing from the constricting orifice.

plasma arc welding (PAW): An arc welding process that uses a constricted arc between a non-consumable electrode and the weld pool or between the electrode and the constricting nozzle. Shielding is obtained from the ionized gas issuing from the torch, and may be supplemented by an auxiliary source of shielding gas.

plug weld: A weld made in a circular hole in one member of a joint fusing that member to another member.

polarity: The condition of being positive or negative with respect to some reference point or object. In welding the terminals of the power supply are designated negative and positive. Which terminal is hooked to the electrode determines polarity.

porosity: A discontinuity formed by gas entrapment during solidification or in a thermal spray deposit.

positive pressure torch: The positive pressure torch requires that gases be delivered at pressures above 2 psi (14 kPa). In the case of acetylene, the pressure should be between 2 and 15 psi (14 to 103 kPa). Oxygen is generally supplied at approximately the same pressure for welding.

post-flow time: The time interval from current shut off to either shielding gas or cooling water shut off.

postheating: The application of heat to an assembly after welding, brazing, soldering, thermal spraying, or thermal cutting.

powder coating: A durable, weather-proof, polymer coating for metals applied in a spray and then cured in an oven.

power factor: The ratio of true power (watts) to the apparent power (volts times amperes). The power factor is equal to the cosine of the angle of lag between the alternating current and voltage wave.

power source: An apparatus for supplying current and voltage suitable for welding, thermal cutting, or thermal spraying.

precipitate: To cause to become insoluble, with heat or a chemical reagent, and separate out from a solution.

precipitation hardening: A multiphase heat treatment process that strengthens alloys by causing phases to precipitate at various temperatures and cooling rates.

preform: Brazing or soldering filler metal fabricated in an application specific-shape or form.

preheat: The heat applied to the base metal or substrate to attain and maintain preheat temperature.

prequalified welding procedure specification: A welding procedure specification that complies with the stipulated conditions of a particular welding code or specification and is therefore acceptable for use under that code or specification without a requirement for qualification testing.

press brake: A hydraulically-powered press used to make sharp bends in heavy plate using mating dies.

pressure regulator: A device designed to maintain a nearly constant supply pressure. Regulators may be attached to pressurized cylinders, gas generators, or pipe lines to reduce pressure as desired to operate equipment.

primary windings: The windings connected to and receiving power from an electrical circuit.

procedure qualification: The demonstration that welds made by a specific procedure can meet prescribed standards.

process: A grouping of basic operational elements used in welding, thermal cutting, brazing, or thermal spraying.

proportional limit: The greatest stress a material can withstand without deviation from the straight-line proportionality between stress and strain.

protective atmosphere: A gas or vacuum envelope surrounding the workpieces, used to prevent or reduce the formation of oxides and other detrimental surface substances, and to facilitate their removal.

PTFE seal: A polyterafluroethylene polymer seal; TEFLON® is the DuPont trademark for PTFE.

puller clips: Steel assembly aids used with wedges and screws to force parts into alignment for welding. They are temporarily welded onto the work and removed after welding is complete.

pulsed-power welding: An arc welding process variation in which the power is cyclically programmed to pulse so that effective but short duration values of power can be utilized. Such short duration values are significantly different from the average value of power. Equivalent terms are pulsed-voltage or pulsed-current welding.

purging: The removing of any unwanted gas or vapor from a container, chamber, hose, torch, or furnace.

push angle: The travel angle when the electrode is pointing in the direction of the weld progression. This angle can also be used to partially define the positions of welding guns.

Q

qualification: A specific set of procedures designed to test a weldor's ability; followed by a weldor qualification test. After passing a particular qualification test a weldor is then qualified to weld to the variables of that qualification.

quenching: The sudden cooling of heated metal by immersion in water, oil, or other liquid. The purpose of quenching is to produce desired strength properties in hardenable steel.

R

reactor: A device used in arc welding circuits to minimize or smooth irregularities in the flow of the welding current; also called an *inductor*.

reducing flame: An oxyfuel flame with an excess of fuel gas.

residual stress: Stress present in a joint member or material that is free of external forces or thermal gradients.

resistance brazing (RB): A brazing process using heat from the resistance to electric current flow in a circuit of which the workpieces are a part.

resistance soldering (RS): A soldering process using heat from the resistance to electric current flow in a circuit of which the workpieces are a part.

resistance welding (RW): A group of welding processes that with the application of pressure produces coalescence of the faying surfaces with the heat obtained from resistance of the workpieces to the flow of the welding current in a circuit of which the workpieces are a part.

resistor: A device with measurable, controllable, or known electrical resistance used in electronic circuits or in arc welding circuits to regulate the arc amperes.

Rockwell hardness test: The most common hardness testing method. This procedure uses a minor load to prevent surface irregularities from affecting results. There are nine different Rockwell hardness tests corresponding to combinations of three penetrators and three loads.

roll brake: A hydraulically powered press which puts smooth bends in plate goods using one round roller to force the workpiece between two other round rollers.

root bead: A weld bead that extends into or includes part or all of the joint root.

root bend test: A test in which the weld root is on the convex surface of a specified bend radius.

root face: That portion of the groove face within the joint root.

root opening: A separation at the joint root between the workpieces.

root pass: The initial welding pass of a multi-pass weld.

root penetration: The distance the weld metal extends into the joint root.

root reinforcement: Weld reinforcement opposite the side from which welding was done.

runoff weld tab: Additional material that extends beyond the end of the joint, on which the weld is terminated.

S

SAE: The Society of Automotive Engineers.

safety disc: A disc in the back side of a high pressure cylinder valve designed to rupture and release gas to the atmosphere preventing cylinder rupture if the cylinder is mishandled.

seal weld: Any weld designed primarily to provide a specific degree of tightness against leakage.

seam weld: A continuous weld made between or upon overlapping members, in which coalescence may start and occur on the faying surfaces, or may have proceeded from the outer surface of one member. The continuous weld may consist of a single weld bead or a series of overlapping spot welds.

semiautomatic: Pertaining to the manual control of a process with equipment that controls one or more of the process conditions automatically.

shear strength: The characteristic of a material to resist shear forces.

shear: To tear or wrench by shearing stress; to cut through using a cold cutting tool when shearing metal.

sheet metal brake: A hand or power tool that puts straight-line bends into sheet metal or plate.

shielding gas: Protective gas used to prevent or reduce atmospheric contamination.

short-circuiting transfer: Metal transfer in which molten metal from a consumable electrode is deposited during repeated short circuits.

side bend test: A test in which the side of a transverse section of the weld is on the convex surface of a specified bend radius.

silicon rectifier: A silicon semiconductor device that acts like a check valve for electricity and is used to change alternating current to direct current.

silicon: A nonmetallic element resembling graphite in appearance, used extensively in alloys. It is the second most common element on earth. Silicon is usually found in the oxide (silicate) form. Silicon contributes to the strength of low-alloy steels and increases hardenability along with performing the valuable function of a deoxidizer, eliminating trapped gas.

single-phase: A generator or circuit in which only one alternating current voltage is produced.

Six F (6F): A welding test position designation for a circumferential fillet weld applied to a joint in pipe, with its axis approximately 45° from horizontal, in which the weld is made in flat, vertical, and overhead welding positions. The pipe remains fixed until welding is completed.

Six G (6G): A welding test position designation for a circumferential groove weld applied to a joint in pipe, with its axis approximately 45° from horizontal, in which the weld is made in the flat, vertical, and overhead welding positions. The pipe remains fixed until welding is completed.

Six GR (6GR): A welding test position designation for a circumferential groove weld applied to a joint in pipe, with its axis approximately 45° from horizontal, in which the weld is made in the flat, vertical, and overhead positions. A restriction ring is added adjacent to the joint to restrict access to the weld. The pipe remains fixed until welding is completed.

slag: A nonmetallic product resulting from the mutual dissolution of flux and nonmetallic impurities in some welding and brazing processes.

slip rolls: A hand- or electrically-powered tool with a set of three or more steel rollers that form sheet metal or plate goods into cylinders or cones by drawing the workpiece into and around the rollers to form it.

slope: A term used to describe the shape of the static volt-ampere curve of a constant-voltage welding machine. Slope is caused by impedance and is usually introduced by adding substantial amounts of inductance to the welding power circuit. As more inductance is added to a welding circuit, there is a steeper slope to the volt-ampere curve. The added inductance limits the available short-circuit current and slows the rate of response of the welding machine to changing arc conditions.

slot weld: A weld made in an elongated hole in one member of a joint fusing that member to another member. The hole may be open at one end.

slugging: The unauthorized addition of metal, such as a length of rod, to a joint before welding or between passes, often resulting in a weld with incomplete fusion.

soft jaws: Plastic, leather, lead, or aluminum covers on the jaws of a vise or pliers used to prevent marking and damage to the work.

solder interface: The interface between solder metal and the base metal in a soldered joint.

solder metal: That portion of a soldered joint that has melted during soldering.

solder: The metal or alloy used as a filler metal in soldering, which has a liquidus not exceeding 840°F (450°C) and below the solidus of the base metal.

soldering iron: A soldering tool having an internally or externally heated metal bit usually made of copper.

soldering: A group of welding processes that produce coalescence of materials by heating them to the soldering temperature and by using a filler metal having a liquidus not exceeding 840°F (450°C) and below the solidus of the base metals. The filler metal is distributed between closely fitted faying surfaces of the joint by capillary action.

solidus: The highest temperature at which a metal or an alloy is completely solid.

solutionizing: The process of dispersing one or more liquid, gaseous, or solid substances in another, usually a liquid, so as to form a homogeneous mixture.

spacer strip: A metal strip or bar prepared for a groove weld and inserted in the joint root to serve as a backing and to maintain the root opening during welding. It can also bridge an exceptionally wide root opening due to poor fit.

spatter: The metal particles expelled during fusion welding that do not form a part of the weld.

spheroidizing: A stress relieving method of long-term heating of high-carbon steel at or near the lower transformation temperature, followed by slow cooling to room temperature.

spliced joint: A joint in which an additional workpiece spans the joint and is welded to each member.

spool: A filler metal package consisting of a continuous length of welding wire in coil form wound on a cylinder (called a barrel) which is flanged at both ends. The flange contains a spindle hole of smaller diameter than the inside diameter of the barrel.

spot weld: A weld made between or upon overlapping members in which coalescence may start and occur on the faying surfaces or may proceed from the outer surface of one member. The weld cross-section is approximately circular.

spray transfer: Metal transfer in which molten metal from a consumable electrode is propelled axially across the arc in small droplets.

stack cutting: Thermal cutting of stacked metal plates arranged so that all the plates are severed by a single cut.

staggered intermittent weld: An intermittent weld on both sides of a joint with the weld increments on one side alternating with respect to those on the other side.

standoff distance: The distance between a welding nozzle and the workpiece.

steel: A material composed primarily of iron, less than 2% carbon, and (in an alloy steel) small percentages of other alloying elements.

step-down transformer: A transformer that reduces the incoming voltage.

step-up transformer: A transformer that increases the incoming voltage.

stickout: In GTAW, the length of the tungsten electrode extending beyond the end of the gas nozzle. In GMAW and FCAW, the length of the unmelted electrode extending beyond the end of the contact tube.

S-to-N curve: A graphical plot showing the incidence of fatigue failure of a part under a given stress level (the S) after N load cycles. These plots are used to determine safe operating stress and part life.

strain: Distortion or deformation of a metal structure due to stress.

stress trajectories: These are diagrams showing the flow of tensile and compressive forces within a beam or structure under load. Only the direction of the forces are shown, not the magnitude.

stress: A force causing or tending to cause deformation in metal. A stress causes strain.

stringer bead: A type of weld bead made without appreciable weaving motion.

stub: The short length of filler metal electrode, welding rod, or brazing rod that remains after its use for welding or brazing.

stud welding (SW): A general term for joining a metal stud or similar part to a workpiece. Welding may be accomplished by arc, resistance, friction, or other process with or without external gas shielding.

submerged arc welding (SAW): An arc welding process that uses an arc or arcs between a bare metal and the weld pool. Molten metal is shielded by a blanket of granular flux on the workpieces. The process is used without pressure and with filler metal from the electrode and sometimes from a supplemental source (welding rod, flux, or metal granules).

substrate: Any material to which a thermal spray deposit is applied.

sulfur: A pale yellow, odorless, brittle, nonmetallic element found underground either in the solid state or as a molten sulfur.

surface preparation: The operation necessary to produce a desired or specified surface condition.

surfacing material: The material that is applied to a base metal or substrate during surfacing.

surfacing weld: A weld applied to a surface, as opposed to making a joint, to obtain desired properties or dimensions.

surfacing: The application by welding, brazing, or thermal spraying of a layer of material to a surface to obtain desired properties or dimensions, as opposed to making a joint.

sweat soldering: A soldering process variation in which workpieces that have been pre-coated with solder are reheated and assembled into a joint without the use of additional solder (also called *sweating*).

swing room: The empty space around one or more pipes needed apply a pipe wrench to a pipe or its fittings, and then to rotate the pipe wrench handle.

T

tack weld: A weld made to hold the parts of a weldment in proper alignment until the final welds are made.

template: Same as a pattern.

tensile strength: The resistance to breaking exhibited by a material when subjected to a pulling stress. Measured in lb/in^2 or kPa.

tension test: A test in which a specimen is loaded in tension until failure occurs.

theoretical throat: The distance from the beginning of the joint root perpendicular to the hypotenuse of the largest right triangle that can be inscribed within the cross-section of a fillet weld. This dimension is based on the assumption that the root opening is equal to zero.

thermal conductivity: The ability of a material to transmit heat.

thermal cutting (TC): A group of cutting processes that severs or removes metal by localized melting, burning, or vaporizing of the workpieces.

thermal expansion: The expansion of materials caused by heat input.

thermal spraying (THSP): A group of processes in which finely divided metallic or nonmetallic surfacing materials are deposited in a molten or semi-molten condition on a substrate to form a thermal spray deposit. The surfacing material may be in the form of powder, rod, cord, or wire.

thermal stress relieving: A process of relieving stresses by uniform heating of a structure or a portion of a structure, followed by uniform cooling.

Three F (3F): A welding test position designation for a linear fillet weld applied to a joint in which the weld is made in the vertical welding position.

Three G (3G): A welding test position designation for a linear groove weld applied to a joint in which the weld is made in the vertical welding position.

three-phase power: A generator or circuit delivering three voltages that are 1/3 of a cycle apart in reaching maximum value. Three-phase current is usually used for circuits of 220 volts or more.

time temperature transformation (TTT): See isothermal transformation diagram definition.

tinning: A non-standard term for pre-coating.

T-joint: A joint between two members located approximately at right angles to each other in the form of a *T*.

torch brazing (TB): A brazing process that uses heat from a fuel gas flame.

torch oscillation: Moving a torch in a back and forth motion.

torch soldering (TS): A soldering process that uses heat from a fuel gas flame.

torsion: The stress produced in a body, such as a rod or wire, by turning or twisting one end while the other is held firm or twisting in the opposite direction.

transferred arc: A plasma arc established between the electrode of the plasma arc torch and the workpiece.

transverse crack: A crack with its major axis oriented approximately perpendicular to the weld axis.

travel angle pipe: The angle of less than 90° between the electrode axis and a line perpendicular to the weld axis at its point of intersection with the extension of the electrode axis, in a plane determined by the electrode axis and a line tangent to the pipe surface at the same point. This angle can also be used to partially define the positions of welding guns, torches, rods, and beams.

travel angle: The angle less than 90° between the electrode axis and a line perpendicular to the weld axis, in a plane determined by the electrode axis and the weld axis. This angle can also be used to partially define the positions of welding guns, torches, rods, and beams.

TSP: The common abbreviation for trisodium phosphate, a strong degreaser and metal cleaner.

tubing saddle: The shape of the end of one pipe as cut to fit snugly against another prior to welding, also called a *fishmouth*.

tungsten electrode: A non-filler metal electrode used in arc welding, arc cutting, and plasma spraying, made principally of tungsten.

Two F (2F) pipe: A welding test position designation for a circumferential fillet weld applied to a joint in pipe, with its axis approximately vertical, in which the weld is made in the horizontal welding position.

Two F (2F) plate: A welding test position designation for a linear fill weld applied to a joint in which the weld is made in the horizontal welding position.

Two FR (2FR): A welding test position designation for a circumferential fillet weld applied to a joint in pipe, with its axis approximately horizontal, in which the weld is made in the horizontal welding position by rotating the pipe about its axis.

Two G (2G) pipe: A welding test position designation for a circumferential groove weld applied to a joint in a pipe, with its axis approximately vertical, in which the weld is made in the horizontal welding position.

Two G (2G) plate: A welding test position designation for a linear groove weld applied to a joint in which the weld is made in the horizontal welding position.

U

U-groove weld: A type of groove weld.

ultimate tensile strength: The maximum tensile stress a material placed in tension can bear without breaking.

under-bead crack: A crack in the heat-affected zone generally not extending to the surface of the base metal.

undercut: A groove melted into the weld face or root surface and extending below the adjacent surface of the base metal.

underfill: A condition in which the weld face or root surface extends below the adjacent surface of the base metal.

uphill: Welding with an upward progression.

V

vertical up: A nonstandard term for uphill welding.

vertical welding position: The welding position in which the weld axis, at the point of welding, is approximately vertical, and the weld face lies in an approximately vertical plane.

V-groove weld: A type of groove weld.

volt: A unit of electrical force or potential.

W

waster plate: A piece of metal used to initiate thermal cutting.

water hammer: The common name for hydraulic shock noise and vibration caused by the rapid deceleration of water in a pipe when a valve closes rapidly.

watt: A unit of electric power equal to voltage multiplied by amperage. One horsepower is equal to 746 watts.

wave soldering (WS): An automatic soldering process where workpieces are passed through a wave of molten solder.

weave bead: A type of weld bead made with transverse oscillation.

weld axis: A line through the length of the weld, perpendicular to and at the geometric center of its cross-section.

weld bead: A weld resulting from a pass.

weld crack: A crack located in the weld metal or heat-affected zone.

weld face: The exposed surface of a weld on the side from which welding was done.

weld groove: A channel in the surface of a workpiece or an opening between two joint members that provides space to contain a weld.

weld interface: The interface between weld metal and base metal in a fusion weld, between base metals in a solid-state weld without filler metal, or between filler metal and base metal in a solid-state weld with filler metal.

weld interval: The total of heat and cool times and upslope time used in making one multiple-impulse weld (resistance welding).

weld joint mismatch: Misalignment of the joint members.

weld metal area: The area of weld metal as measured on the cross-section of a weld.

weld metal: The portion of a fusion weld that has been completely melted during welding.

weld pass sequence: The order in which the weld passes are made.

weld pass: A single progression of welding along a joint. The result of a pass is a weld bead or layer.

weld penetration: A nonstandard term for joint penetration and root penetration.

weld pool: The localized volume of molten metal in a weld prior to its solidification as a weld metal.

weld puddle: A nonstandard term for weld pool.

weld reinforcement: Weld metal in excess of the quantity required to fill a joint.

weld root: The points, shown in a cross-section, at which the root surface intersects the base metal surfaces.

weld symbol: A graphical character connected to the welding symbol indicating the type of weld.

weld tab: Additional material that extends beyond either end of the joint, on which the weld is started or terminated.

weld toe: The junction of the weld face and the base material.

weld: A localized coalescence of metal or nonmetals produced either by heating the materials to the welding temperature, with or without the application of pressure, or by the application of pressure alone, with or without the use of filler material.

weldability: The capacity of material to be welded under imposed fabrication conditions into a specific suitably designed structure and to perform satisfactorily in the intended service.

welder: Weldor is the preferred spelling. for one who performs manual or semi-automatic welding, as opposed to welder being a machine used for welding.

welding arc: A controlled electrical discharge between the electrode and the workpiece that is formed and sustained by the establishment of a gaseous, conductive medium called an arc plasma.

welding electrode: A component of the welding circuit through which current is conducted and that terminates at the arc, molten conductive slag, or base metal.

welding filler metal: The metal or alloy to be added in making a weld joint that alloys with the base metal to form weld metal in a fusion welded joint.

welding helmet: A device equipped with a filter plate designed to be worn on the head to protect eyes, face, and neck from arc radiation, radiated heat, spatter, or other harmful matter expelled during some welding and cutting processes.

welding leads: The workpiece lead (cables) and electrode lead (cables) of an arc welding circuit.

welding operator: One who operates adaptive control, automatic, mechanized, or robotic welding equipment.

welding positions: The relationship between the weld pool, joint, joint members, and welding heat source during welding.

welding power source: An apparatus for supplying current and voltage suitable for welding.

welding procedure qualification record (WPQR): A record of welding variables used to produce an acceptable test weldment and the results of tests conducted on the weldment of a qualified welding procedure specification.

welding procedure specification (WPS): A document providing the required welding variables for a specific application to assure repeatability by properly trained weldors and welding operators.

welding procedure: The detailed methods and practices involved in the production of a weldment.

welding rod: A form of welding filler metal, normally packaged in straight lengths, that does not conduct the welding current.

welding schedule: A written statement, usually in tabular form, specifying values of parameters and welding sequence for performing a welding operation.

welding sequence: The order of making welds in a weldment.

welding symbol: A graphical representation of a weld.

welding wire: A form of welding filler metal, normally packaged as coils or spools, that may or may not conduct electrical current depending upon the welding process with which it is used.

welding: A joining process that produces coalescence of materials by heating them to the welding temperature with or without the application of pressure, or by the application of pressure alone with or without the use of filler metal.

weldment: An assembly whose component parts are joined by welding.

weldor performance qualification: The demonstration of a weldor's ability to produce welds meeting prescribed standards.

weldor's chalk: A soft, naturally-occurring soap stone used by weldors for marking metal.

wetting: The phenomenon whereby a liquid filler metal or flux spreads and adheres in a thin continuous layer on a solid base metal.

wiped joint: A joint made with solder having a wide melting range and with the heat supplied by the molten solder poured onto the joint. The solder is manipulated with a hand-held cloth or paddle to obtain the required size and contour.

wire feed speed: The rate at which wire is consumed in arc cutting, thermal spraying, or welding.

work angle: The angle less than 90° between a line perpendicular to the major work-piece surface and a plane determined by the electrode axis and the weld axis. In a T-joint or a corner joint the line is perpendicular to the non-butting member. This angle can also be used to partially define the positions of guns, torches, rods, and beams.

work hardening: Also called cold working; the process of forming, bending, or hammering a metal well below the melting point to improve strength and hardness.

work-piece lead: The electrical conductor between the arc welding current source and work-piece connection.

workpiece: The part that is welded, brazed, soldered, thermal cut, or thermal sprayed.

wrap-around: A leather, paper, or plastic pipe-fitting tool that is wrapped around the outside of a pipe to locate the minimum-circumference line for a square cut.

wrinkle bending: A bending process using oxyfuel torch heat to put a series of small bends or wrinkles in the wall of a pipe. Together these wrinkles can form a much larger bend.

wrought iron: A material composed almost entirely of iron, with very little or no carbon.

Y

yield strength: The load at which a material will begin to yield, or permanently deform. Also referred to as yield point.

yield strength: The stress at the uppermost point on the straight-line portion of the stress-strain curve. Stress imposed on the sample below this level produces no permanent lengthening and stress can vary from zero up to the yield strength. Stress above yield strength causes permanent lengthening.

Young's modulus: A ratio between the stress applied and the resulting elastic strain; the slope of a metal's elastic limit curve; a relative measure of a material's stiffness. Also known as the modulus of elasticity.

Index

3-4-5 triangle, 29
Adding threads by welding, 191
Assembly, product, 174
Base plate designs, 215
Beach marks, 281
Beam,
    deflection, 273-275
    cambering, 125-127, 139
    drilling through, 269
    expanded, 213-215
    stress distribution in, 267-271
    -to-column connections, 216
Bearing,
    supports, 199
    removing stuck, 155-156
Bending,
    machines, 105-114
    equipment, 102 (table)
    flame, 119-122, 124-125
    mandrel, 116-117
    shrinking, 119-122, 135-137
    straightening, 119-122, 135
    circles, 107
    zero-radius, 108
    scrolls, 109
    coils and springs, 110
    tubing, 111
    angle iron, 112
    channels, 113
    squares, 114
    wrinkle, 117
Bessey®,
    corner clamps, 8
    welding clamps, 7
Bolt,
    removal methods, 153-154
    parts of, 243

    snap-off, 245
Bracing, diagonal, 162-163
Brackets,
    195
    angle iron, 34
    welded, 201-202
Brake,
    press, 104-105
    roll, 105
    sheet metal, 104
    Burning bars – see Exothermic burning bars
Burn-through,
    of tubing, 150-151
    vehicle sheet metal, 147-148
Cambering I-beams, 125-127
Castings, brazing of, 249-250
Checklist, welding design, 160-176
Chill bars, 182
Clamping work at right angles, 160
Clamps,
    angle for tubing, 8
    corner, 8
    DE-STA-CO Industries, 11, 16
    general purpose, 6-7
    Vise-Grip®, 9-10
    welding for pipe, 12
    welding, 7
    welding platen, 19, 21
Cleats, 198
Column splices, 216-217
Columns,
    279
    attaching to and guying, 217
Concrete,
    prestressed, 275-279
    reinforcing, 275

strength of, 275
Corner tools, magnetic, 30
Crack, stopping propagation, 156
Cutting problems, oxyfuel,
    224-225
Degreasing agents, 24
Di-Acro,
    bender, 106
    table top slip rolls, 119
Distortion,
    controlling, 176
    causes of weld, 178
Drill rod, brazing of, 248-249
Edge preparation of pipe, 50-51
Elastic matching, 203
Elasticity, common metals,
    260 (chart)
Elasticity, modulus of, 258,
    258 (chart)
Exothermic burning bars, 229-234, 230
    (chart)
Extrusions, bending, 106
Fabrication,
    hand tools, 3
    power tools, 5-6
    steps, 2
Fasteners,
    aircraft, 239-240
    commercial, 235
    metric, 240, 240 (chart), 241 (chart),
        242 (chart)
    pretensioning, 244-246
    SAE, 235-236 (chart)
    structural steel guidelines, 243
    structural steel, 237, 237 (chart), 238
        (chart), 239 (chart)
    threaded, 234-246
    threaded, types, 234 (chart)
Fatigue,
    failure fracture interfaces, 282
    life, factors, 286
    life, impact of grain orientation on,
        292-293
    life, measuring, 286-289
    life, modifications to increase, 290-
        291
Fixture, pin, 103
Fixtures, 14-16
Flame,
    bending, 119-126

bending rolled shapes,
    124-125
shrinking, 137-138
straightening, 120
Flap wheel, 25
Forged steel pipe fittings, 64-65
Frame,
    repairs on vehicles, 141-146
    box, 33
    mitered, 27
    notching, 28
    rectangular, 27-28
    building square, 28-29
    squaring, 31
    torsional resistance of, 162
Hangars for rolled shapes, 220
Hasps, 193
Hinges, 24
Hooke's law, 253
I-beams,
    205-208
    cutting aids, 227-228
    cutting of, 227
Indicators, Direct Tension (DTIs), 244-
    245
Insert rings, 55-56
Intermittent welds, chain and staggered,
    182
Jigs, bending, 103
Joint,
    design, welded, 170-173
    fittings, 63-64
Lances, oxygen, 229-234
Lead-free solders, 78
Leveling jacks, 191
Line heating, 127-134
Machine bases, 199-200
Mechanical properties, definitions of, 252
Metal,
    cleaning preparation, 24-25
    degreasing, 24-25
    fatigue, 280-292
    fatigue, beginning of, 281
    markers, 5
    marking methods, 4
    protective finishes, 25-26
Miters, structural pipe, 222
Moment of inertia, 272-273
Mounting plates on columns,
    218-219

Noise and vibration, controlling, 158-159
Nut, retaining of, 197
Nuts, cutting off with oxyfuel, 228-229
Overwelding, 180
Panels, stiffening, 158
Parts layout, 164-169
Patterns, 14
Pins, adding, 193
Pipe,
    alignment tools, 52-55
    manufacturing processes, 42-43
    Master template maker, 148
    versus tubing, 35-36
    welding advantages, 43-33
Piping codes, 44 (chart)
    cutting circle in, 226-227
    piercing of, 225-226
    preparation of for welding, 169-170
Pliers, sheet metal, 103
Postweld heat treatment, 62-63
Preheating pipe for welding, 56
Problems
    with cover plates, 138-139
    with fences and railings, 137-138
Puller clips, 13
Pyramids, cutting 18
Rings,
    making from plate, 166
    nested segments of, 166
Roll bars, mounting, 147
Rolled
    sections, steel, 208-210
    shapes, commercially available, 11-115
Rolls, slip and bender, 118-119
Root pass grinding, 59
Sand filling, for pipe, 118
Sealing up closed vessels, 151
Section, adding plates to, 221-222
Shafts,
    building up worn, 154-155
    stress distribution in, 277-278
Shrinkage and distortion, 181-184
    controlling factors, 179
Shrinkage,
    longitudinal in beams, 185-186
    transverse and longitudinal, 179
    transverse, calculating, 184-185
Slag removal, 58
Snugging, 243
Solder consumption on joints, 86, 87 (chart)
Squares, 3
Steel,
    available shapes and forms, 21-22
    physical property changes with temperature, 177 (graph)
    sizes available, 23
Stiffeners, vertical, 162
S-to-N curves, 286-288
Strain, calculations of, 256-257
Strength, compressive, 266-267
Stress,
    concentration factor, 284
    concentrations, causes of, 282-283
    concentrators, 285
    trajectories, 269-270
    allowable, 265-266
    calculations of, 256
    examples of, 255
    sudden, 279-280
    types of, 254
    -strain curves, 262-265
Strong backs, 183
Subassemblies, use of, 174
Table legs, attaching, 31-32
Table, design of
    cutting, 17
    welding, 17
Tack welds on pipe, 57-58
Temperature indicating crayons, 129
Tensile testing, 261-262
Thickness, adding to a panel, 157
Threads, adding to tubing, 197
Tie-downs, deck, 198
Tips, brazing and soldering, 246-250
Tools,
    hand, 3-4
    power, 4-5
Torch,
    holder, 246-247
    pressures, oxyfuel, 223, 224 (chart)
Trammel points, 4
Truss design, 271
Tubing,
    closing ends of, 194-195

temporary repair of copper, 89-90
Twist, resisting of steel shapes, 161-162
Unibody, mounting roll bars on, 147
Van Sant template maker, 148
Weld,
- access holes, 217-218
- size guidelines, 188-198
- size, 173-174
- strength, 190-119
- cleaning and inspection, 176

Welded pads, securing, 157
Welding,
- galvanized or cadmium plated metal, 151
- machine, extending capacity of, 150
- platens, 19-21
- positions, 47
- procedures, 175-176
- qualification, 45
- rebar, 220
- thick to thin parts, 150
- work environment, 2
- angle iron, 204-205
- screening on frames, 152-153

Welds,
- cracking, 212-213
- hooking, 210-212
- measuring fillets, 186-188
- parts of, 45-46

Young's modulus – see elasticity
Zero stress points, locating welds at, 204

Credits

American Beauty – Assembly Technologies International, Inc.
Figure 3-43.

Applied Bolting Technology Products, Inc.
Figures 6-116, 6-117.

Chicago Metal Rolled Products, Inc.
Figure 4-16.

Copper Development Association Inc.
Tables 3-6, 3-7; Figure 3-53.

Di-Acro, Incorporated
Figures 4-7, 4-8, 4-9, 4-10, 4-11, 4-12, 4-13, 4-14, 4-22.

DE-STA-CO Industries
Figures 1-13, 1-14, 1-19.

Haydon Bolts, Inc.
Figure 6-118.

James F. Lincoln Arc Welding Foundation Publisher of *Design of Weldments* and *Design of Welded Structures* by Omar Blodgett.
Figures 4-37, 4-38, 6-8, 6-9, 6-10, 6-11, 6-12, 6-14, 6-15, 6-16, 6-17, 6-18, 6-19, 6-20, 6-21, 6-22, 6-23, 6-24, 6-25, 6-26, 6-27, 6-28, 6-29, 6-30, 6-31, 6-32, 6-33, 6-34, 6-35, 6-36, 6-37, 6-38, 6-39, 6-40, 6-41, 6-42, 6-45, 6-46, 6-47, 6-49, 6-50, 6-51, 6-52, 6-53, 6-54, 6-55, 6-56, 6-57, 6-58, 6-70, 6-71, 6-72, 6-73, 6-74, 6-75, 6-76, 6-77, 6-78, 6-79, 6-80, 6-87, 6-88, 6-89, 6-90, 6-91, 6-92, 6-93, 6-94, 6-95, 6-96, 6-97, 6-98, 6-99, 6-100, 6-103, 6-104, 6-105, 6-106 (upper pair), 7-3, 7-7, 7-9, 7-10, 7-24, 7-25, 7-30, 7-35, 7-36, 7-37; Tables 6-1, 7-3; Welding Design Checklist.

Mathey Dearman, Inc.
Figure 3-14.

MK Products Inc.
Figures 3-54, 3-55.

Nucor Corporation, Fastener Division
Table 6-7, 6-8.

Oxylance Corporation
Table 6-3; Figure 6-114 (torch handle).

Van Sant Enterprises Inc.
Figures 4-17, 5-8.

U.S. Bureau of Land Management, *Metric Handbook*
www.blm.gov/nhp/efoia/wo/handbook/handbook.html
Tables 6-10, 6–11.

Victualic Company of America
Figure 3-62.

Vise-Grip® – American Tool Companies, Inc.
Figures 1-9, 1-10, 1-11, 1-12, 4-1.

Walhonde Tools Inc.
Figures 3-15, 3-16, 3-17.

Weldsale Company
Figures 1-22, 1-23, 1-24.